MATV SYSTEMS HANDBOOK

Design, Installation & Maintenance

By Allen Pawlowski

FIRST EDITION

FIRST PRINTING—JULY 1973

Printed in the United States
of America

Hardbound Edition: International Standard Book No. 0-8306-3657-9

Paperbound Edition: International Standard Book No. 0-8306-2657-3

Library of Congress Card Number: 73-78204

Preface

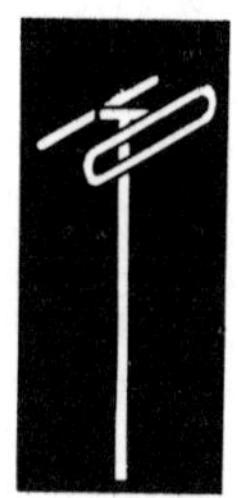

Chances are that you probably have a master antenna television system (MATV) right in your own home. You may not have thought of it that way but just consider this accepted definition:
"An MATV system delivers television signals to two or more TV sets from a common or master antenna." This implies that a home antenna feeding TV sets in both the bedroom and living room is by definition an MATV system. Such systems range in complexity from two-set installations in a private home to networks with hundreds and even thousands of outlets for TV in apartment buildings, motels, and hospitals, schools, mobile home parks, and on and on.

The first MATV system probably was born in some apartment building in New York City when the second tenant in that building bought a TV set and found that the landlord wouldn't let him put another antenna on the roof. Undaunted, he went to his neighbor across the hall and asked to hook onto his antenna. When this didn't work out too well, he bought an antenna signal booster that probably improved reception on both TV sets. As more and more tenants got TV sets, they bought more and more boosters—**and a new business was born!**

This modest beginning actually created two industries. The first was MATV, which satisfied the needs of apartment dwellers in the city. The second business started about the same time but in outlying areas. This business is today's CATV industry with millions of people receiving their television via cable. Of these two industries, CATV is widely known through frequent articles and exposure in the news. MATV is also a major industry but goes along almost unnoticed because it serves a noncontroversial need of the TV viewing public.

The aim of this book is to explore the **hows** and **whys** of MATV system design, installation, and maintenance. This requires as much attention to the **hows** or practical considerations as it does to the **whys** or theoretical considerations. In fact, the practical and theoretical are so interdependent they are discussed throughout this book with no serious attempt to separate them. It is my strong belief that the installation technician is a great asset to his company when he has an appreciation for some of the theoretical aspects of MATV. Similarly, and probably more important, an MATV system designer needs to have a feel for the practical side of system work before he can design good systems.

I assumed in this book that you might have no experience with MATV whatever, but that you are familiar with TV transmission and reception. If you **are** in MATV, you will find the first few chapters an illuminating review. For both the pro and the novice, I have deliberately included material that affects your **economic** well being—in particular, ways and means to increase the dollor-producing capabilities of old systems and how to design new sources of income into old or new systems. Of course, I've also pointed out economic and technical pitfalls.

Although this book is primarily concerned with MATV, you will find the design end applicable to CATV systems as well. The strong foundation developed from this text will serve you well in understanding the special considerations of CATV not covered here.

Various manufacturers' products are illustrated as examples of the type of units discussed in this book. No endorsement of any product is intended or implied. However, we gratefully acknowledge the help received from the following manufacturers who contributed product information.

Blonder-Tongue Laboratories, Inc., Old Bridge, N.J.
Channel Master Corp., Ellenville, N.Y.
The Finney Company, Bedford, Ohio
Jerrold Electronics Corp., Horsham, Pa.
Winegard Company, Burlington, Iowa.

Allen Pawlowski

Contents

Basics

All professions have their jargon. For doctors it's centered around drug chemistry, and an ability to write unreadable prescriptions. Electronics is no different. Many specialized names and terms are needed to describe what goes on in electronic circuits and systems. Common to all electronics and vital to the comminications industry is an understanding of **decibels**. Decibels are used to describe the performance of products such as antennas, amplifiers, cable, etc., as well as signal strength, loss, and a host of other items regarding signal quality. A good understanding of decibels is a very powerful tool in MATV. With this understanding comes the ability to make intelligent selections of products and designs of systems that work.

DECIBELS

Decibels are, in reality, very common to everyday life. The only reason they are not known to everyone is that the subject isn't usually included in high school physics. An example of the decibel is seen every time you turn on a lamp containing a three-way bulb. Let's say you have a lamp with a 50-100-150 watt bulb. At the 50-watt setting you see a certain amount of light output. Switching to the 100-watt setting, you notice a rather large increase in the amount of light, about double. Changing to the 150-watt setting gives an increase but a much smaller **apparent** one, something less than 50 percent. We all know that at each setting the power was increased by 50 watts. This is known as a **linear** change. Our eyes, however, did not see a linear or equal change but rather they responded to the **percentage** of change. For us to have seen equal increases in light, the bulb settings would have to have been 50, 100, and 200 watts.

This follows a **logarithmic** rate of change rather than a linear rate of change. Decibels are used in electronics to express the logarithmic changes of signal in a system, needed to produce apparently equal changes to the eye and ear. If you have a working knowledge of logarithms, decibels will be easy. Decibels are still easy to understand without this knowledge but a review would never hurt. For the mathematically minded the following formula for decibels (dB) relates to MATV work.

$$dB = 20 \log_{10} \frac{V2}{V1}$$

Where:

dB is the ratio of voltage change
20 is a constant
$\log_{10}$ is the common log to the base 10
V1 is the voltage input
V2 is the voltage output.

Two important things can be seen in this formula. First, the decibel is a logarithmic function. Second, the decibel is a ratio between two voltages. The voltage formula is used because of the practical consideration of measuring equipment. In MATV systems, signals are measured with a **field-strength meter** (FSM) sometimes known as **signal intensity meter**. In reality, these are sensitive, tunable voltmeters. In broadcasting and some other electronic specialty services, power meters are used to measure signal strength.

EXAMPLES OF DECIBEL USE

Because there is more than one way to measure signal strength, confusion often occurs when relating signal level changes to decibels. This happens when you state the size of signal level changes as a ratio. For instance, a 2:1 increase **in power** is a 3 dB change; however, 6 dB represents a 2:1 increase **in voltage** or **current**. Although this may be confusing at first, rest assured that there is no difference in the size of the decibel. This can be seen by looking at what happens when the signal level changes across a resistor, or constant antenna circuit impedance. A doubled voltage increases a signal level by 6 dB. Since the resistance stays constant, this can only occur if the current through the resistor also doubles. A doubling of both the voltage **and** current produces a 4:1 increase in power. Since a 4:1 is brought about by doubling the voltage, either case represents a 6 dB change.

It is important to understand that using decibels is a method of expressing the size of a change. To make working with the decibel easier, I recommend that you commit to memory two decibel relationships. They are: 6 dB equals a **voltage** ratio of 2:1, and 20 dB equals a **voltage** ratio of 10:1. There is a table of decibel equivalents in the appendix which lists the voltage ratios from 1 to 60 dB. Since such a table will not always be handy, remembering the ratios for 6 dB and 20 dB will serve you well for most practical purposes.

To illustrate this better, let's assume that you have an amplifier with 6 dB of gain. This says that whatever signal is applied to the input, the amplifier will increase signal strength by a factor of 2 at the output. If the input were 500 microvolts, the output will be 1000 microvolts. If the input were 4000 microvolts, the output will be 8000 microvolts. From this you can see that the gain of the amplifier is independent of actual signal level, assuming the signals are within the normal range of operation for that amplifier.

An amplifier with a gain of 20 dB will have the ability to increase the strength of any signal by a factor of 10. In this case, an input signal level of 693 microvolts would appear at the output at a level of 6930 microvolts. Again the amplifier gain is independent of signal strength.

Now let's put these two memorized decibel factors to work. Assume next an amplifier with a gain of 26 dB, as in Fig. 1-1. What is the gain factor? To make this easy you can think of this amplifier as having two transistors, one giving 6 dB of gain and the other, 20 dB. The 6 dB section would double the signal and 20 dB would increase that level by 10 times. Thus the total increase will be 2 x 10 or 20 times. If a signal level of 200 microvolts is applied to the input, the first transistor will double that level. This 400 microvolt signal is then applied to the second transistor which will raise its level by 10 times to 4000 microvolts. Thus it can be seen that 26 dB is simply a combination of the two familiar factors, added together.

Another example would be an amplifier with 32 dB of gain. Here you can simply treat this as having three stages, a 6, a 6 and a 20 added together to get 32. But remember that the

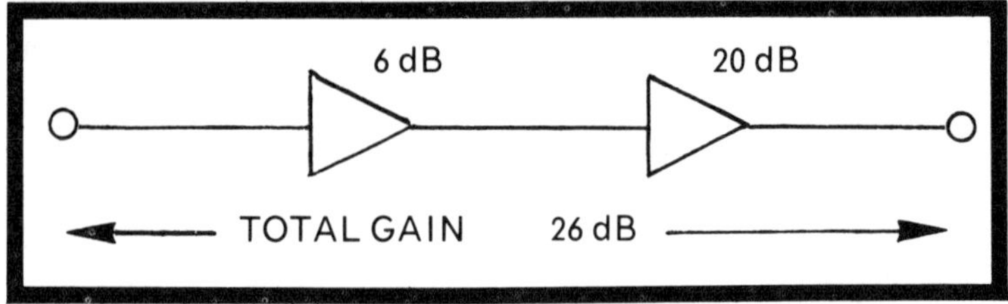

Fig. 1-1. An amplifier with 26 dB of gain can be looked at as being two amplifiers in cascade, one having 6 dB of gain and the other with 20 dB of gain.

effects **multiply**; the first stage doubles any signal, the second stage doubles that signal for a subtotal of 4, and the third stage raises that signal by 10 times for a total **amplification** of 40 times.

Getting a little more complicated, let's consider an amplifier with only 14 dB of gain, as shown in Fig. 1-2. Here we change our thinking a little to consider this an amplifier with two stages, one having 20 dB of gain and one having 6 dB of **loss**.

The amplification factor for this example can be looked at as being 10 for the 20 dB section and ½ for the minus 6 dB section, or an overall amplification factor of 5. This example shows that decibels are subtractive as well as additive.

A thorough understanding of the last two examples will give you the necessary foundation for the full use of decibels. As you become more familiar with them, you will see that they make the job of system design a simple matter of addition and subtraction. To help cement this understanding it is a good idea to go through a few more examples of decibels vs amplification factor.

60 dB gain = 20+20+20 or
10x10x10 = 1000

52 dB gain = 20+20+6+6 or
10x10x2x2 = 400

54 dB gain = 20+20+20-6 or
10x10x10x½ = 500

28 dB gain = 20+20-6-6 or
10x10x½x½ or 25.

It can be seen by trying a few examples of your own that you cannot determine the exact factor for every decibel

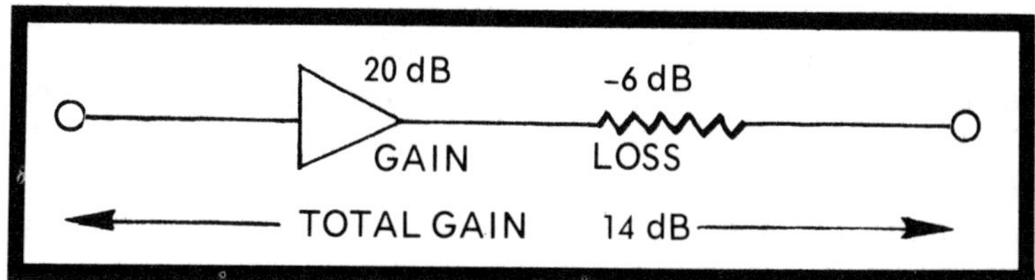

Fig. 1-2. A 14 dB gain amplifier illustrated as a 20 dB gain amplifier followed by 6 dB of attenuation.

number from the factors memorized. You can come very close, however, and in most cases your estimate will be more than adequate. For example, estimate the factor for a 31 dB gain amplifier. The closest you can estimate is:

30 dB = 6+6+6+6+6 or
2x2x2x2x2 = 32 times.

32 dB = 20+6+6 or
10x2x2 = 40 times.

Therefore 31 dB can be estimated to have a gain factor of 36, which is sufficiently close to the actual factor of 35 times.

Not all devices have gain. Cable, for instance, has **loss**. You can apply the same logic to losses as you do to gain. If a certain type of cable has a characteristic loss of 2 dB per 100 ft, then 300 ft of this cable would have 6 dB of loss. This says that whatever signal strength is fed into one end will arrive at the other end at half the input voltage level. Similarly, if the cable were 1000 ft long, it would have 20 dB of loss. Signal strength at the end of this length would be one-tenth the level at the input.

USING DECIBELS TO DESCRIBE SIGNAL LEVELS

We have seen that decibels can be added and subtracted. It is important to remember that the decibel always refers to a ratio of change. Decibels can also be used to describe actual signal levels. This is done by agreeing on some standard reference level. In recent years there has been widespread agreement throughout the communications industry on using 1 millivolt (0.001 volt) as a reference. In MATV also, 1 millivolt across 75 ohms is accepted as the decibel reference for signal level measurements.

Unfortunately the term **millivolt** (mV) has never become popular in the TV industry. Most common is the use of **microvolts**, abbreviated **uV**, when measuring signal level. The transformation between millivolts and microvolts is easy in that 1000 uV is equal to 1 mV. Any level given in millivolts can be converted to microvolts by simply adding three zeros to the number. Should you desire to convert microvolts to millivolts, just move the decimal point three places to the left.

A requirement of using decibels to describe actual signal levels is to identify the difference between decibels meaning a ratio of change and decibels meaning a signal strength. **This is done by writing "dBmV" whenever you refer to an actual signal level.** The dB part says you are using the decibel notation and the mV part means you are using the 1 mV as a reference. One item is always assumed in MATV work when using **dBmV**. It is assumed that the circuit impedance is always 75 ohms. The term dBmV is never used when talking about signal levels in a 300-ohm circuit. In MATV work this is no handicap since all coaxial cable (coax) and equipment used is designed for 75-ohm operation. The only exception to this is where the system connects to TV sets or antennas having 300-ohm terminals. Table 2 in the appendix is a full listing of the microvolt equivalents for all dBmV values from minus 40 to plus 80 dBmV.

The real value of converting all signal levels to their reference level (dBmV) equivalents shows itself when you start to trace signals through an MATV system. A couple of examples will illustrate this point.

It has been stated that 1000 uV is equivalent to 0 dBmV. If a 0 dBmV signal level were applied to an amplifier with 20 dB of gain, the output signal strength would be plus 20 dBmV or 20 dB stronger than the input. This is determined by adding the amplifier gain to the 0 dBmV input signal level. Since we know that 20 dB is a multiplication factor of 10, then we also know that the output signal level must be 10,000 uV. If the 0 dBmV input signal were applied to a run of cable with minus 20 dB **loss**, the output signal strength would be minus 20 dBmV. We know that 20 dB loss is a factor of 0.1; therefore minus 20 dBmV is 0.1 times the input level, or 100 uV.

Let us now apply this knowledge to a few practical examples to gain some familiarity with the application of the terms **dB** and **dBmV** to systems.

Fig. 1-3 illustrates a home type of antenna and preamplifier (preamp) installation serving one TV receiver. Here the objective is to determine the signal strength delivered to the TV receiver. The following procedure is the basis of all MATV system caluculations.

The antenna signal strength is given as 100 uV. This must first be converted to the proper dBmV equivalent. It can be seen that 100 uV is one-tenth the reference level of 1000 uV and therefore must be 20 dB **below** the 0 dBmV reference, or minus 20 dBmV. This signal strength is increased by the preamp gain of 20 dB:

–20 dBmV +20 dB = 0 dBmV at the preamp output.

This signal then goes through 6 dB of cable loss before reaching the TV. Its level therefore is:

0 dBmV-6 dB = -6 dBmV.

We know that minus 6 dB is a factor of 50 percent; therefore, minus 6 dBmV must be half the reference level. So a 500 uV signal is delivered to the TV tuner.

The next problem, Fig. 1-4, is similar to the above except that two sets are connected to one antenna and the signal levels are different. Again the antenna signal of 200 uV must be converted to dBmV. In the first example we found that 100 uV = minus 20 dBmV. Since 200 uV is 6 dB greater than 100 uV it must be–20 dBmV+6 dB=–14 dBmV. The cable loss after the signal split is the same for both TV sets. In this case it is

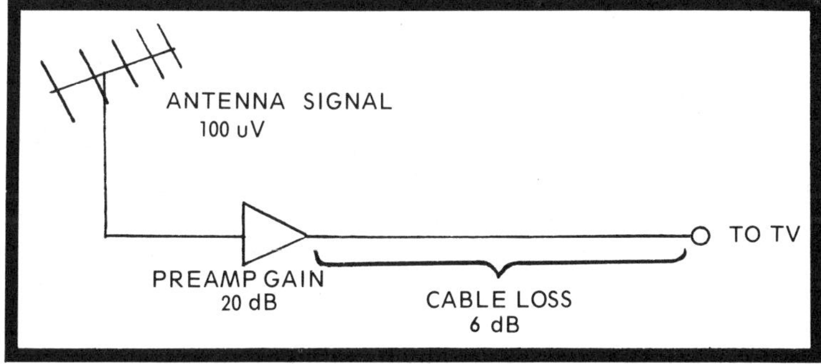

Fig. 1-3. Basic system diagram of amplified signals from an antenna. Text answers the question of how much signal is delivered to the TV receiver in dBmV and microvolts.

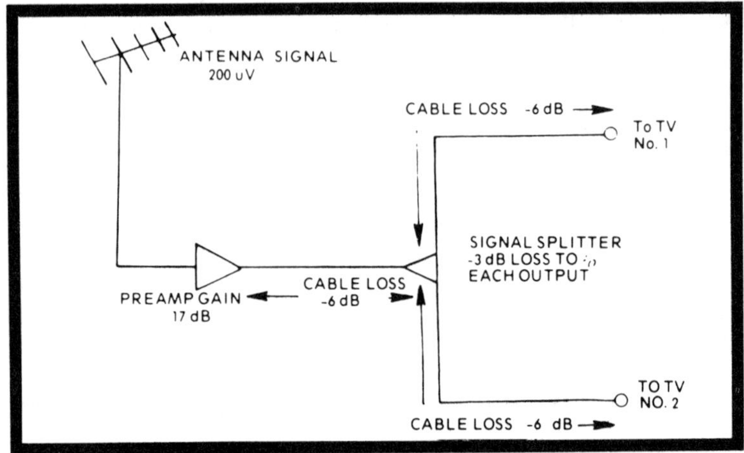

Fig. 1-4. Two-outlet MATV systems. Text explains how to calculate signal strength delivered to TV sets.

safe to assume that if you calculate the level for one set you will get the same results for the second set. In larger systems it is recommended to tabulate your calculations. This problem is tabulated as follows:

− 14 dBmV	Antenna signal level, 200 uV
+ 17 dB	Preamp gain
+ 3 d BmV	Signal strength at preamp output
− 6 dB	Cable loss to splitter
− 3 dBmV	Signal strength to splitter input
− 3 dB	Splitter loss to each output
− 6 dBmV	Signal level at splitter output
− 6 dB	Cable loss, splitter to TV
−12 dBmV	Signal level to TV set

The signal level to each TV is −12 dBmV as tabulated above. Converting this level back to microvolts we see that −12 dBmV is the same as (−6 dB plus −6 dB) below the 1000 uV reference. This would be ¼ the reference level or 250 uV to each receiver.

So far we have discussed examples that might be found in home antenna installations. The next example will entail an MATV system as may be installed in a small motel. See Fig. 1-5. It is suggested that you try to determine the signal strength delivered to the TV outlet illustrated. Test yourself! Use the same procedure of tabulation as in the previous example. Do not get confused by the variety of devices used in these

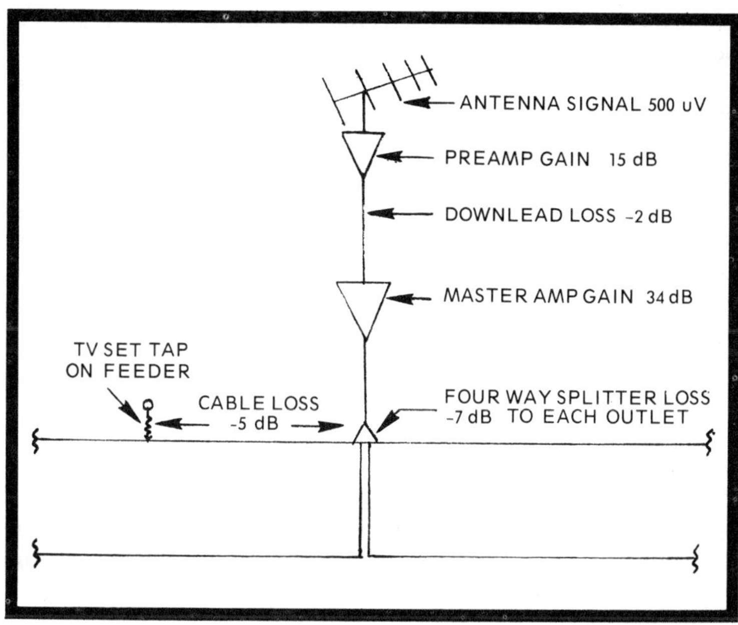

Fig. 1-5. Partial diagram for an MATV system in a small motel. Text explains simplified method of determining signal strength delivered to TV set after accounting for all devices in the system.

examples as their functions will be explained in Chapter 2. This illustration is a simplified one showing only a single tap to one receiver. Such a system could have many connections to supply signal to each room.

Now that you've tried your hand, check your calculations against mine:

− 6 dBmV	Antenna signal level
+ 15 dB	Preamp gain
+ 9 dBmV	Preamp outlet
− 2 dB	Downlead loss
+ 7 dBmV	Input level to master amplifier
+ 34 dBmV	Master amplifier gain
+ 41 dBmV	Master amplifier output
− 7 dB	Four-way splitter loss
+ 34 dBmV	Output level to each feeder line
− 5 dB	Cable loss to tap
+ 29 dBmV	Feeder cable signal strength at tap location
− 29 dB	Tap loss
0 dBmV	Signal level to each TV set

Since 0 dBmV is equal to the reference level, then the level of signal delivered to the TV is 1000 uV, or 1 mV.

These examples have shown how to calculate signal strength into TV sets from known levels of signal at the antenna. The purpose of these exercises is to gain familiarity with manipulating **dB** and **dBmV**. As we have seen, it's simply a matter of adding and subtracting gains and losses to the dBmV equivalent of signal level. If you would like to confirm this simplified approach, use Fig. 1-5 again and try to determine the signal strength at the receiver using **microvolts of signal** at each point in the system.

Many other applications of decibels in system engineering present themselves. We shall explore a few more just to sharpen your ability to handle these terms. What size amplifier would you purchase or specify for the system illustrated in Fig. 1-6? Here the problem is turned around somewhat. Since we show how much signal is required at the TV set, it is necessary to work **backward** to determine the amplifier gain requirements. Study the following tabulation carefully.

+ 6 dBmV	Signal level required at TV set
—(—17)dB	Loss of feeder line tap
+ 23 dBmV	Signal level required in feeder cable to overcome tap loss
—(—6)dB	Loss of cable
+ 29 dBmV	Signal level required to overcome cable loss
—(—7)dB	Four-way splittter loss
+ 36 dBmV	Signal level required at amplifier output to overcome **all** losses.

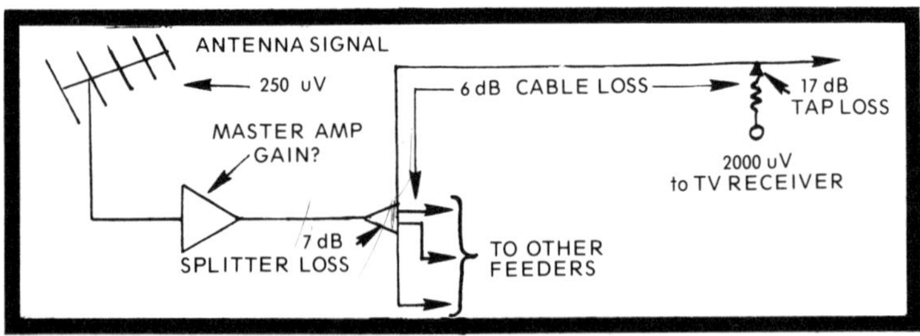

Fig. 1-6. Partial MATV system diagram. Text answers the question of unknown master amplifier gain.

Now that we know how much signal is needed to drive the system, this can be compared with the signal available from the antenna, to derive the master amplifier gain.

+36 dBmV	Required signal strength at amplifier output
−(−12) dBmV	Signal strength available from the antenna
+ 48 dB	Gain required in master amplifier

One final example should help to illustrate the value of system calculations using decibels. The problem here is to determine the proper value of loss for the attenuator pad when connecting a closed-circuit camera into the system of Fig. 1-7.

Many self-contained closed-circuit cameras are available. Most have the ability to deliver an rf output that can be carried over an MATV system. For such a system to be successful it is necessary that the signal strength of the closed-circuit channel be approximately equal to the level of all other channels in the system. This is called **balancing the system**. If this is not done, the closed-circuit channel delivered to the TV sets may be too weak and therefore, snowy. If the camera signal is too strong, as in this example, the amplifier will be overloaded and interference to all channels will occur.

The answer to this problem is readily found by the tabulation method, starting from some known point.

+ 48 dBmV	Master amplifier output level
—(+42) dB	Amplifier gain
+ 6 dBmV	Master amplifier input level
—(—8) dB	Directional coupler mixing loss
+ 14 dBmV	Signal needed to overcome mixing loss.

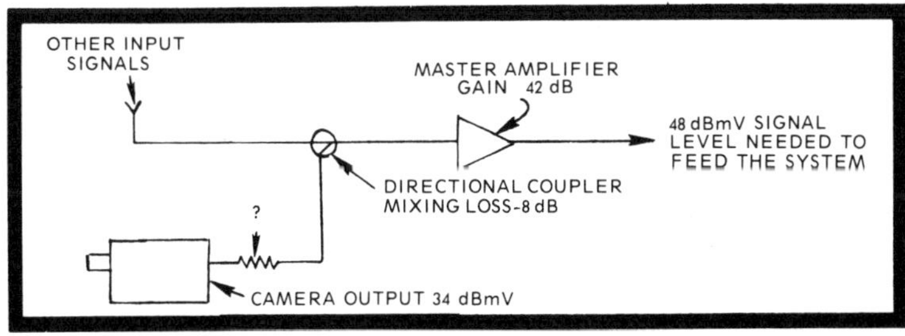

Fig. 1-7. Illustration of typical closed circuit camera connection to an MATV system. Text explores the question of proper attenuator value to properly balance all signal levels in the system.

Fig. 1-7 shows a camera output level of 34 dBmV available. In the tabulation above we find that a 14 dBmV signal is needed to properly drive the master amplifier, through the mixing device. Comparing 34 dBmV with 14 dBmV tells you to insert an attenuator pad of minus 20 dB (**loss**) to balance the signals in the system.

HOW MUCH SIGNAL IS ENOUGH

Signals of various levels are found throughout every TV distribution system. There is a need to know what to expect at various levels of operation. This is very important to an MATV system designer because it is his responsibility to design a system that will supply snow-free signals to all sets. Installation and maintenance technicians should also know this; how else would they judge when the job of installation or repair is complete?

The objective here is to **assure** snow-free operation of the TV sets connected to the system. Snow is **electrical noise**, very similar to the hiss of a radio when tuned between stations. Snow-free operation means that there is enough signal present to drown out the snow thus producing a clear, sharp picture. We can therefore say **the amount of signal that drowns out the snow is enough**. This is considered minimum. Every TV receiver also has a maximum. This too is valuable knowledge since too much signal can produce even worse effects than too little.

Signal-to-Noise Ratio

We are referring to the quality of signal which is known as **signal-to-noise ratio** (SNR). The noise is always with us. Even in apparently snow-free pictures, there is a small amount of noise, but fortunately it is so weak compared to the signal that our eyes can't see it. The noise present is largely generated by the tubes and transistors in the electronic circuits of amplifiers and TV receivers. A small amount of additional noise is given to us by nature. All these sources of noise are caused by the random motion of electrons because the parts are warm. This is often called **thermal noise**. Tubes generate a lot of noise because they run quite hot. Transistors also run hot because of the current flowing through them. Since transistors run cooler than tubes, they generally produce less noise.

To find a starting point it is necessary to imagine an amplifier that is perfect. If this were possible, the only noise present would be that noise given to us by nature. Since an amplifier doesn't know the difference between TV signals and random noise, we would actually measure noise at the output of this perfect device. See Fig. 1-8. The 75-ohm resistor used to terminate the amplifier input with a simulated load is really a noise generator. It is actually the lowest-noise-producing element operating at room temperature.

The noise generated by a 75-ohm resistive source and delivered to the input of a properly matched amplifier is 1.1 uV when considering the normal 4.2 MHz bandwidth of a TV picture. This bandwidth is used in these calculations because we are primarily interested in the noise or snow that appears on a regular TV receiver. We will therefore assume this bandwidth in all further discussions about noise. Translating 1.1 uV, it appears at a level of −59 dBmV. If the amplifier in Fig. 1-8 had a gain of 30 dB, the noise meter connected to the output would measure a noise level of −29 dBmV when restricted to a 4.2 MHz bandwidth.

Let us now examine what happens with real amplifiers. Examine Fig. 1-9. Using a typical amplifier with 30 dB gain in place of the theoretically perfect one of the previous example, we find that the noise meter reads a noise level 6 dB stronger. The discrepancy is attributable to the amplifier's **noise figure**.

With transistors it is not unusual to find noise figures ranging between 2 and 15 dB. Preamps located close to the antenna are primarily optimized for best low-noise operation. Preamps of 2 to 8 dB are common depending on price and frequency of operation. Amplifiers with higher noise figures are generally designed for high-signal-level operation; thus, the small percentage of noise contributed can be ignored.

Using the numbers in Fig. 1-9, let us work backward from the output. The noise meter reads −23 dBmV. If we subtract the gain of the amplifier, 30 dB, we get −53 dBmV. This is an

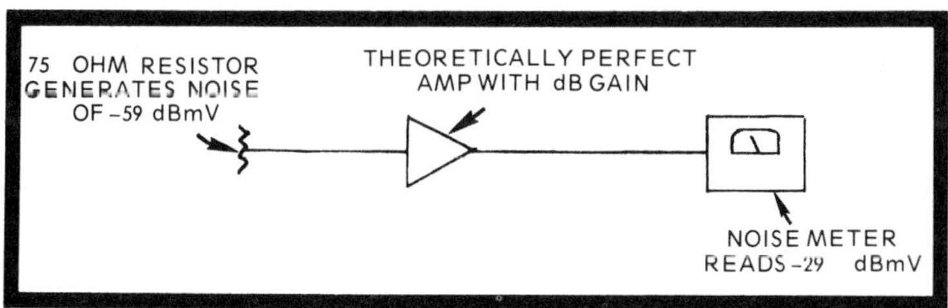

Fig. 1-8. A theoretically perfect amplifier will amplify the thermally generated noise developed in the resistor used to terminate the input.

imaginary number known as **equivalent noise input** That is to say, if the amplifier were perfectly noise-free, the only way the noise meter could read −23 dBmV on the output would be to insert a noise level of −53 dBmV into the input. From this point we can start to analyze picture quality.

The important parameter in determining minimum usable signal level is the SNR. If no signal is present then all we see on the TV is noise or snow. If equal amounts of noise and signal are present, we have a ratio of 1 to 1, or 0 dB SNR. If as explained above, the equivalent noise input is −53 dBmV and the actual signal level input is −13 dBmV then the SNR is 40 dB or (10 x 10) is **100 to 1**.

Tests of SNR

Simply playing around with numbers will not give us the answer to how much signal is needed for snow-free operation. This problem was recognized in the very early days of telecasting, along with a host of other unknowns. To gain a better understanding, the Federal Communications Commission authorized a study group to look into the problems associated with TV broadcasting and reception. The group was known as the Television Allocations Study Organization, or **TASO**. In their report submitted to the FCC in 1959 were the results of subjective testing of SNR.

The test involved the viewing of a standard TV receiver by many people under controlled conditions. This test was made many times. For each test the SNR was adjusted to different levels and the people were asked to rate the pictures accordingly. Remember that these tests were conducted when most telecasting was black and white. Television today must cope with the rigors of color transmissions and it has been found necessary to slightly increase the SNR vs picture quality ratings given in the TASO report. Table 1-1 is the TASO report updated with consideration for color.

Table 1-1. Picture quality rating vs SNR (Adjusted for color).

SNR, dB	Subjective Picture Quality Rating
45	Excellent. A picture of extremely high quality. As good as could be desired.
40	Good. A picture of good quality providing enjoyable viewing. Interference perceptible.

SNR, dB	Subjective Picture Quality Rating
33	Passable. A picture of acceptable quality. Interference is not objectionable.
28	Marginal. A picture of poor quality with somewhat objectionable interference.
22	Poor. A just-watchable picture with definitely objectionable interference.
Below 22	Unusable. Pictures too bad for viewing.

The information in Table 1-1 may now be coupled with knowledge about natural noise and noise figures. We can set goals for MATV system signal levels that will produce excellent quality television. We can also predict, with some degree of certainty, the quality of a TV picture if we know the signal strength.

Let us review the example in Fig. 1-9. We calculated that this amplifier has an equivalent noise input of –53 dBmV. If the objective were to provide **excellent**, or snow-free quality pictures, the minimum signal level into this amplifier must be 45 dB stronger than the noise. That is –53 dBmV + 45 dB = –8 dBmV. Converting, this becomes 400 uV. In MATV work, the porrest quality which you should accept is the rating **good**. This would be for **this** amplifier –53dBmV + 40 dB = –13 dBmV or approximately 225 uV.

A brief word of MATV philosophy is in order. When you contract to design, install, or operate an MATV distribution system, your performance is based primarily on the quality of

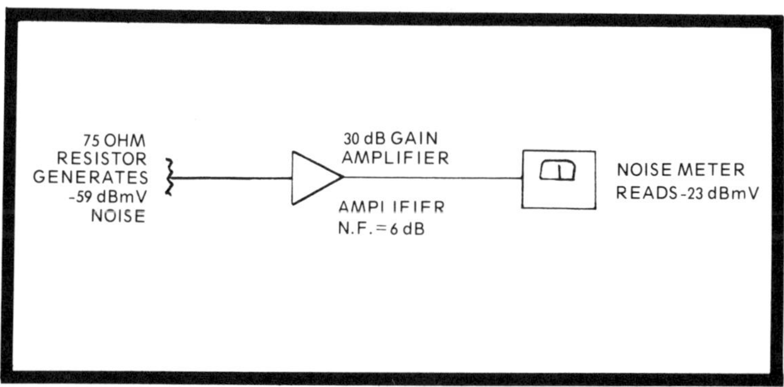

Fig. 1-9. Noise power meter is used to measure the noise output of an amplifier. From this the amplifier's noise figure can be determined.

the pictures on the TV set. It is strongly recommended that you set your goal sufficiently high to achieve that **excellent** rating. If you want to get paid for the job, there can be no argument with this kind of performance. If you were paying someone for the job, you certainly wouldn't want to accept anything less.

As in all engineering, there are exceptions to the basic rules. Just enough signal to qualify for snow-free operation is generally not enough. Your responsibility for a job does not necessarily end the day the check clears the bank. All electronic devices change with age. The area from which you can expect the most trouble is the TV receivers connected to the system. As sets get older, they generally lose sensitivity. In engineering terms it means that their noise figures increase. Under these degraded conditions it takes more and more signal to produce a snow-free picture. On the other hand, more signal to each receiver costs extra because amplifiers with better boost are needed in the system.

It is recommended that the minimum signal level delivered to any receiver be at least 1 mV, or 0 dBmV. The average new TV of today has a noise figure around 8 dB; therefore, minimum signal for snow-free operation calculates at:

−59 dBmV	Basic natural noise
8 dB	Average noise figure
−51 dBmV	Equivalent noise input
45 dB	Minimum SNR objective
−6 dBmV	Minimum signal input to reach objective
6 dB	Safety or maintenance margin
0 dBmV	Recommended MATV system output level to each TV.

The safety margin of 6 dB shown in the calculation is recommended as minimum in all MATV system work. The cost of this amount of margin is usually small and will more than pay for itself if it prevents a false alarm service call the first time someone's TV hiccups.

Another reason to deviate from this ideal condition comes up in tough fringe-area situations. Depending upon location and the desires of your customer, you may find a picture rated **marginal** is the best picture in town. Fringe-area reception, discussed in Chapter 4, can occur even in large metropolitan areas where sports-minded viewers attempt reception from distant stations during locally blacked out events.

Results of such efforts can be favorable if the approach is logical. By understanding minimum signal requirements you can judge your ability to satisfy a customer. More importantly, you can determine when it's time to quit a lost cause.

In such instances, it is first desirable to get as much signal as possible. This calls for a big antenna. To preserve the quality of signal received, you next select a preamp with a low noise figure. To assure best performance, you locate the preamp as close to the antenna as practical to avoid losing very weak signals in lossy transmission lines.

Again minimum signal level can be calculated. Let us assume that you are willing to live with a **marginal** picture. Assume also that a preamp with a 3 dB noise figure is not prohibitively expensive. Minimum acceptable signal under these conditions is:

–59 dBmV	Basic natural noise
3 dB	Preamp noise figure
–56 dBmV	Equivalent noise input
28 dB	Marginal quality SNR
–28 dBmV	Minimum usable signal.

A signal level of –28 dBmV is equal to 40 uV and will be quite snowy; **but if the customer wants it bad enough, it could be a profitable job.**

A good working knowledge of the basic applications of decibels is essential to success in MATV engineering. The material covered in the following chapters will draw heavily on subjects discussed here. As larger systems are discussed, the tabulations of results will sometimes get quite lengthy but the level of mathematical complication will not go beyond addition and subtraction of positive and negative numbers.

Selecting Equipment

All equipment used in MATV is advertised by way of performance specifications (specs). This holds true for the smallest connector or the largest electronic unit. In general, the basic methods of measuring performance specs are uniform throughout the industry. There are a few important specs, however, which have no widespread standard and manufacturers are free to set their own standards.

This chapter will take a look at the various elements that make up an MATV system. We will explore the specs relative to each item in an attempt to gain an understanding of their meaning.

The aim of this discussion is to familiarize you with the variety of product specs encountered in the field. An understanding of these specs will serve you well in selecting and evaluating the products of all suppliers in your goal to design and build high quality, reliable MATV systems.

ANTENNAS

Of all the products used to make up an MATV distribution system, the antenna is the most important single item, for it is the main source of signals carried on the system. The quality of these signals must be good to start with since no amount of doctoring after the antenna can make up for poor original quality.

You are probably already familiar with the different types of antennas available for consumer use. In MATV work it is desirable to use a better grade of antenna than for home TV. Primarily the difference between home and commercial antennas is **durability**. Commercial antennas are generally constructed using seamless aluminim tubing of at least ⅜-inch diameter. Rust resistant bolts are used instead of rivets, and clamps are heavy duty to fit up to 2-inch or larger diameter pipe.

Antennas so constructed tend to cost more; but since they will serve many TVs, their cost per outlet is really quite small. The fact that they serve so many people is sufficient reason to install the best available. Failure of an MATV antenna will produce many complaints and could cause a very costly repair.

Electronically, MATV antennas are of three major designs—**Yagis**, **log periodics**, and **aperture** types. Yagis characteristically have high gain and narrow bandwidth. This makes them best suited for single-channel reception. Log periodics generally have moderate gain but wide bandwidth. This makes them suitable for reception of all channels from one general direction at moderate distances from the transmitters. The third type of antenna includes the various forms of a dipole and reflective screen that is commonly used for reception of UHF. They range in shape from full **parabolic reflectors** through **parabolic sections** to **corner reflectors**. All come under the general heading of **aperture antennas**. Fig. 2-1 shows various examples of these three types.

The important specifications in selecting antennas for MATV work are: **gain**, **bandwidth**, **match**, **polar pattern**, and **mechanical considerations**.

Gain and Bandwidth

In television work, gain of an antenna is specified in decibels above a reference dipole. The dipole is used as a reference since it is the easiest to construct and provides accurate, predictable results every time. If a specific antenna is quoted as having 6 dB of gain, it means that this particular antenna will deliver 6 dB more signal than a dipole under similar conditions. Exactly how much gain is needed to do a particular job is discussed in Chapter 4. Generally speaking, the further away from the transmitter, the more gain needed in the receiving antenna. One curious thing about antenna gain is that the antenna can give you increases in signal strength without increasing the noise. This is the only element in an MATV system with that ability.

Antennas are offered in a variety of bandwidths. Yagis or other single-channel antennas are tuned to perform best at one specific TV channel. Difficulties involving bandwidth can easily occur on the low-frequency TV channels, especially

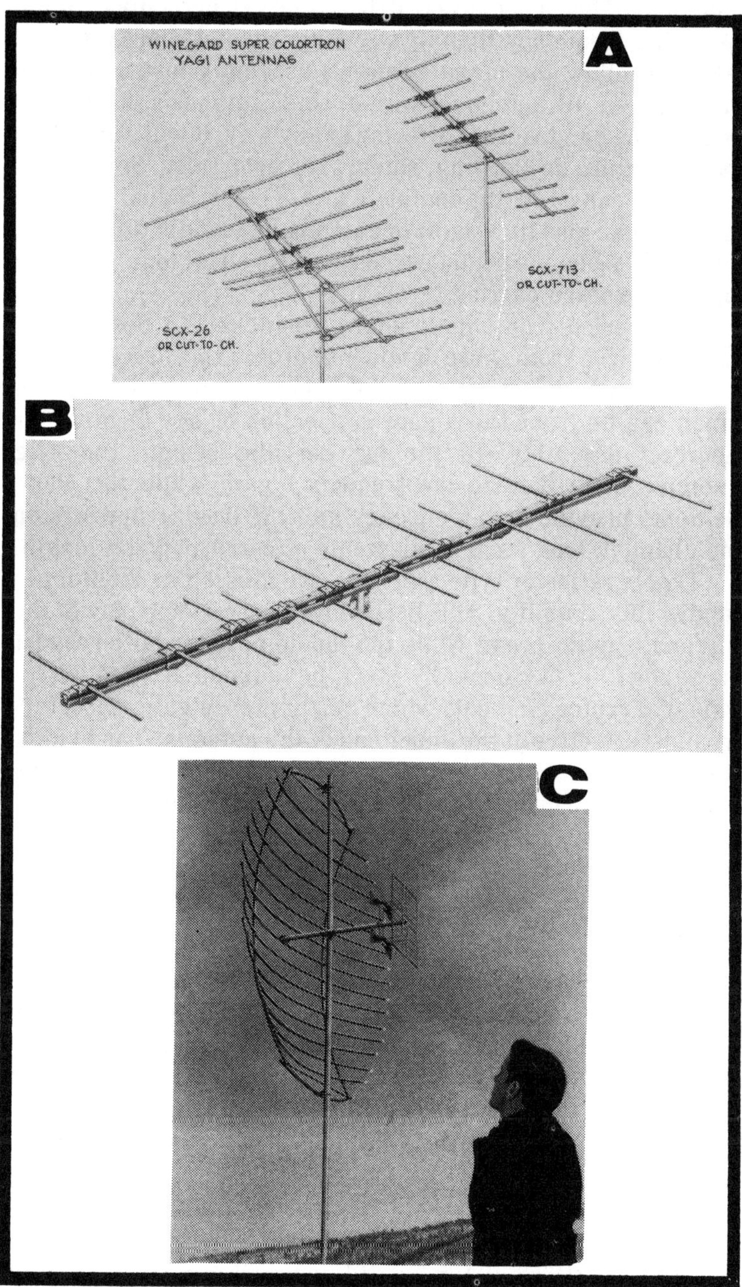

Fig. 2-1. Three types of antennas. A. Yagi, model SCX courtesy Winegard. B. Log periodic, model J-105-Hi courtesy Jerrold. C. Parabolic, model 4250, courtesy Channel Master.

Channel 2. This is because 6 MHz is a larger percentage of the operating frequency than at any other chanel. Care must be taken to follow the manufacturer's assembly instructions to the letter on all antennas, with special emphasis on the low band. If the bandwidth is affected in any way, it will show up in the TV picture as ghosting, smears, or poor color. Sometimes this effect can cause the complete loss of color. Signal strength readings can also tip you off to antenna bandwidth difficulties, especially when the sound carrier level is equal to or stronger than the picture carrier level.

Log periodic antennas, being broadband in design, are generally free from these bandwidth difficulties. Occasionally a log periodic antenna will exhibit low gain on one channel which can be traced to a poor connection of one element, or incorrect assembly. In the log periodic design, the long elements contribute to low-frequency gain while the short elements provide high-frequency gain. If trouble appears on one channel, look for it in elements of corresponding length.

Dipole reflector type antennas are classed as medium in bandwidth capability. This is because the reflector part of the antenna is quite broad while the dipole part tends to restrict the bandwidth. Occasionally this type antenna is fitted with a train of director elements which help increase gain. Directors are most effective at the upper end of the antennas bandwidth. Most MATV manufacturers have two or more models of this type antenna in order to cover the entire UHF band.

Impedance Matching

Match implies an impedance and in TV work, 300 ohms and 75 ohms are standard. A **balun**, a balanced-to-unbalanced transformer, or ordinary matching transformer is required to match 300-ohm antennas to 75-ohm MATV equipment. Match is a very important antenna spec. For years, match has been stated as a **voltage standing wave ratio (VSWR)**. In recent years, match specifications have been appearing in decibels. Graph 4 in the appendix is a very helpful comparison of VSWR to dB match.

Match is best understood when considering the case of a transmitting antenna. The objective is to radiate all the power of the transmitter into the air. If the antenna were perfect, all the energy sent to it would be radiated. Since antennas are not perfect, they exhibit what is called **mismatch** and reflect back some of the energy sent to them from the transmitter. The amount of energy reflected can be measured with a special bridge or directional coupler.

If the voltage level of the reflected energy were half the level of energy sent to the antenna, it would be said that the antenna had a 6 dB match, not too good as antennas go. If the voltage level of the reflected energy were one-tenth the transmitted level, the antenna would have a 20 dB match. This would be considered quite good.

Stated in simple terms, the match of an antenna is its ability to accept and radiate signals. The amplitude of the signal reflected back down the line will be lower than the upgoing signal by the stated match in decibels.

In MATV work, our antenna will be used for receiving signals instead of transmitting them. Match, however, is not limited to antennas; it is an important spec for all parts of a system. Consider what would happen if an antenna with a 6 dB match were connected to a preamp with a 6 dB match. See Fig. 2-2. Signals received by the antenna would travel down the lead-in and enter the preamp. Since it is poorly matched, some of this signal will be reflected back toward the antenna. This reflected signal would be 6 dB weaker than the received signal strength. When this reflected signal reaches the antenna, some of it would be radiated but another reflection would occur. This second reflection would be down another 6 dB. We now have a condition where a signal is headed toward the preamp that is 12 dB weaker than the originally received signal. Since this doubly reflected signal has made one

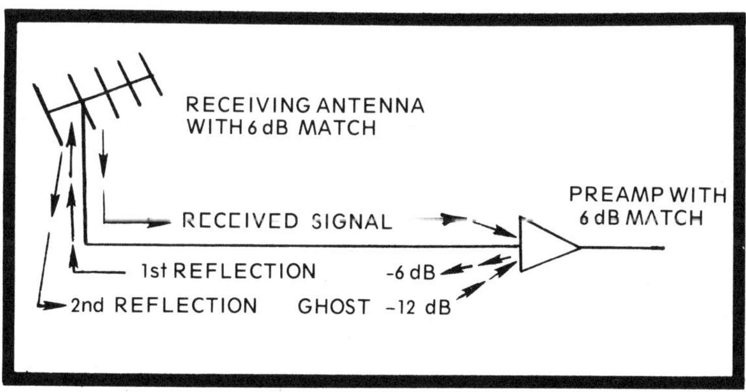

Fig. 2-2. Ghosting can be caused by reflections from poorly matched antenna and preamp terminals.

complete round trip, down and up and back down the transmission line, it is slightly later in time than the original signal that entered the antenna. This delayed signal will appear on the TV set as a ghost in the picture. The longer the down-lead cable, the wider spaced the ghost. It is therefore important to use well matched equipment in MATV to assure ghost-free operation. It has been found that matches of 14 dB are about minimum to meet this objective.

Antennas are made to favor signals from one direction. Unfortunately they do not completely reject signals from other than the forward direction. The measurement of how well an antenna receives or rejects signals cannot be defined by a single term. The polar pattern shows the antenna's receiving ability for all angles.

Your knowledge of decibels vs voltage ratios is put to the test in reading polar patterns because of the way the patterns are made. Typically the antenna is mounted on a rotatable platform something like a home rotor installation. With the antenna aimed toward a transmitter, the received signal voltage level is adjusted to be maximum as plotted on circular or polar graph paper. The antenna is then rotated through a full 360 deg and the signal voltage received is plotted for all angles around the antenna. Fig. 2-3 is a polar plot for the antenna pictured in Fig. 2-1B.

Mixers, Filters, and Traps

The need for passive devices occurs in almost every MATV antenna installation. Very seldom will it happen that a single antenna receives all available channels with interference-free pictures at the desired level. Having individual antennas for each channel gives the system designer greater flexibility toward achieving desired results. When treated separately, each channel can be filtered clean of interference. High Q traps can be inserted to eliminate stubborn interference. After satisfactorily clean pictures are achieved, the signals must be mixed for amplification and distribution. Most manufacturers offer products that perform the filtering and mixing process all in one unit. Fig. 2-4 illustrates a 3-channel-plus-FM mixing filter for VHF and a 3-channel mixing filter for UHF.

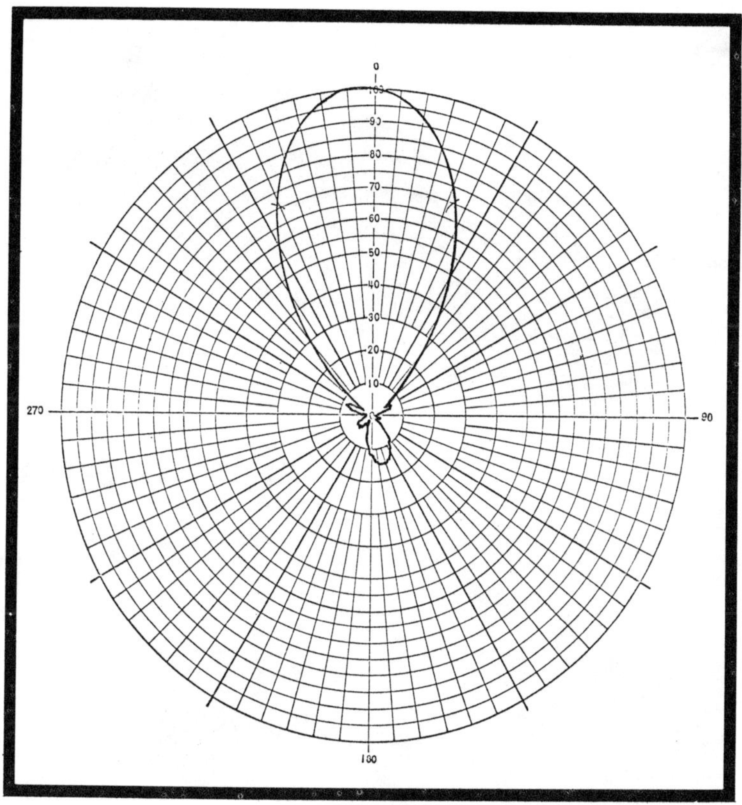

Fig. 2-3. Polar pattern for model J-105-Hi antenna shown in Fig. 2-1B.

The most common need for filters occurs when separate antennas are used for each channel. Antennas in this case are tuned to the proper channels but they do not completely reject the signals of other nearby channels. Assume the need to receive Ch. 7 and Ch. 9. Assume also that the signals come from two different directions, thus demanding separate Yagi antennas. From the previous discussion on polar patterns, we found that even at extreme side angles the antenna will receive some signal.

Fig. 2-5 illustrates the correct and incorrect method of combining these antennas. In 'A' the antennas are simply combined by using a hybrid splitter as a mixer. Any Ch. 9 signal picked up by the Ch. 7 antenna will join with the desired signal from the Ch. 9 antenna in the mixer. This unwanted signal coming through the wrong antenna will be severely distorted by the poor off-channel characteristics of that

Fig. 2-4. Antenna mixing filters. A. Three-channel plus FM mixing filter for VHF low band. Note that this unit includes plug-in attenuator pads for each channel for ease of balancing channel levels. (Model ME-26 courtesy Winegard.) B. Three-channel mixing filter for UHF channels. Unit employs cavity tuned filter circuits. (Model UMN-3, courtesy Jerrold.)

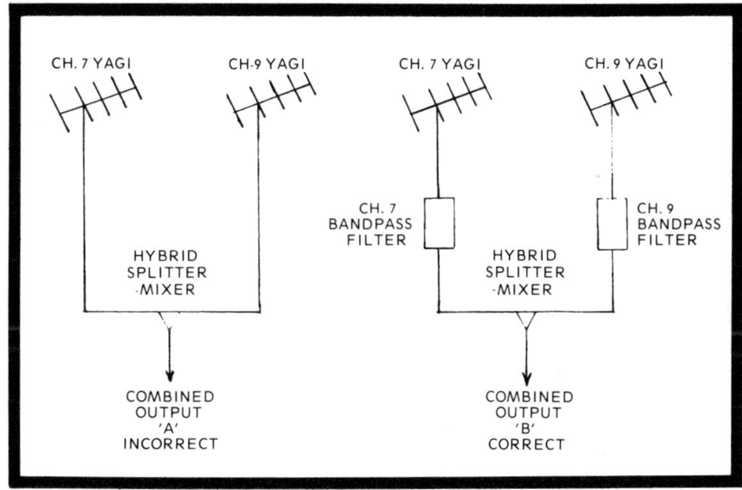

Fig. 2-5. Incorrect (A) and correct (B) method of mixing individual Yagi antennas. Antennas must be filtered to eliminate unwanted reception by wrong antenna.

antenna. The resulting picture quality after mixing antenna signals will be degraded. In 'B' the output of each antenna is passed through a filter to eliminate everything but the desired signal from reaching the mixer. This way original signal quality is maintained.

In general most mixing devices are combinations of various filter elements. A commonly used item is the VHF-UHF mixer. Here the mixer is a combination of a low-pass filter with a high-pass filter. Fig. 2-6 illustrates the action of this type mixing unit. The low-pass section is arranged to insert minimum loss to VHF Ch. 2 through 13. Above Ch. 13 the loss increases sharply until it reaches about minus 30 dB. This says that any UHF signals coming in the VHF terminal will be attenuated some minus 30 dB before reaching the common output. The reverse holds true for the high-pass section.

This device permits the use of separate VHF and UHF antennas. It combines these signals with minimum loss and no worry about unwanted distortion. Note that for products like those discussed, the thing that limits them to signal mixers is the name on the label. Mixing filters can be used equally as well as signal splitters. The split made would be a frequency split. In the case of single-channel combining networks, if signals from a broadband antenna were put into the common terminal, each channel would be separated out by its ap-

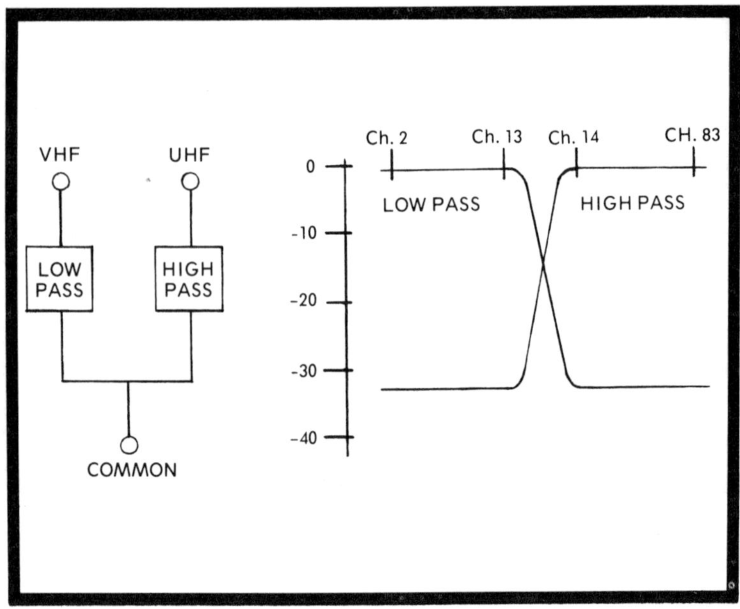

Fig. 2-6. Combination low pass and high pass filters used to combine or separate the UHF and VHF bands, with typical response curve.

propriate filter. Similarly the UHF-VHF mixer could find application as a splitter to separate the bands from a common line for application to the separate input terminals of a TV set.

The important specifications for mixers, filters, and traps are as follows:

Insertion loss is the loss to the desired signal as it passes through the device. Be sure that it is taken into account when tabulating system losses. High quality devices are generally limited to under 1.5 dB on VHF channels and under 3 dB on UHF channels.

Match is an ever-present criterion to good MATV design. Good match in filters is a sign of quality. Poor match can produce ghosting in the same manner as described earlier. Since match is difficult to achieve in these items, it becomes economically sensible to accept somewhat reduced numbers with minimums in the 9 to 10 dB area.

Selectivity is the prime function of a trap or filter. It is a measure of how well unwanted signals are rejected. Interference traps should provide over –40 dB of single-frequency rejection. Special-purpose traps with less rejection will be discussed elsewhere. Single-channel filters come in many varieties. For antenna mixing, selectivity is usually measured at plus and minus 9 MHz from channel center frequency. This tells how well it rejects semiadjacent channels, such as Ch. 7 from Ch. 9. For local applications, 12 to 15 dB selectivity is considered adequate. Where weak, distant signals are sought in the presence of strong local signals, special filters giving 40 dB or more are available.

AMPLIFIERS AND PREAMPS

Amplifiers for MATV work are available in such a wide variety that it is difficult to keep up with them. Fortunately there are categories into which amplifiers fall that can be considered separately. Within each category are subcategories that further simplify understanding. All MATV amplifiers have one thing in common, however; they process signals related to TV broadcasting.

Fig. 2-7 illustrates the assignment of frequencies throughout the VHF and UHF bands. Also shown is an expanded view of a typical TV channel. Study this illustration to become familiar with the various frequencies involved. It would be a good idea to memorize the more important frequencies such as band edges, the Ch. 4 to Ch. 5 gap, the FM band, and the different frequencies within the typical TV channel. The biggest single value of knowing these frequencies is in hunting out sources of interference. For this reason it is important to know what services are assigned to frequencies outside the TV band since these are frequently the source of such troubles.

Preamplifiers are a general category of amplifier usually associated with antennas and weak signals. For this reason preamps are designed for low-noise operation and are usually packaged for outdoor use. The two major types of preamps are **broadband** and **single-channel**. Units shown in Fig. 2-8 are typical of MATV system preamps.

Broadband preamps are available in many types, low VHF, high VHF, all VHF, partial UHF, full UHF, and all-channel. Not all broadband preamps are alike even though they may cover the same number of channels. Some contain a single amplifier that covers the entire band. Others are made up of two or three amplifiers in parallel, each tuned to a different portion of the TV spectrum. The specifications will give you a clue as to what's inside. Let's review these specs to see what they really mean.

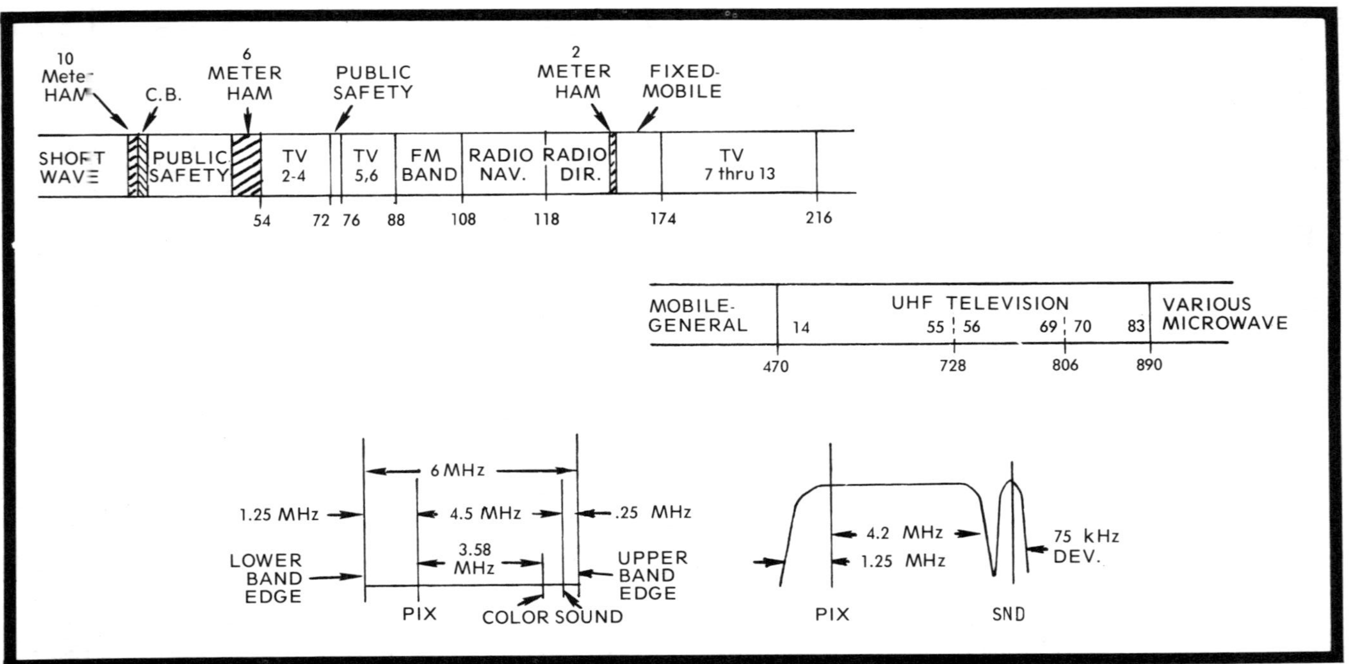

Fig. 2-7. Generalized allocation of frequencies in and around the VHF and UHF television bands. Location of carriers for a standard TV channel and extent of modulation within a standard TV channel assignment.

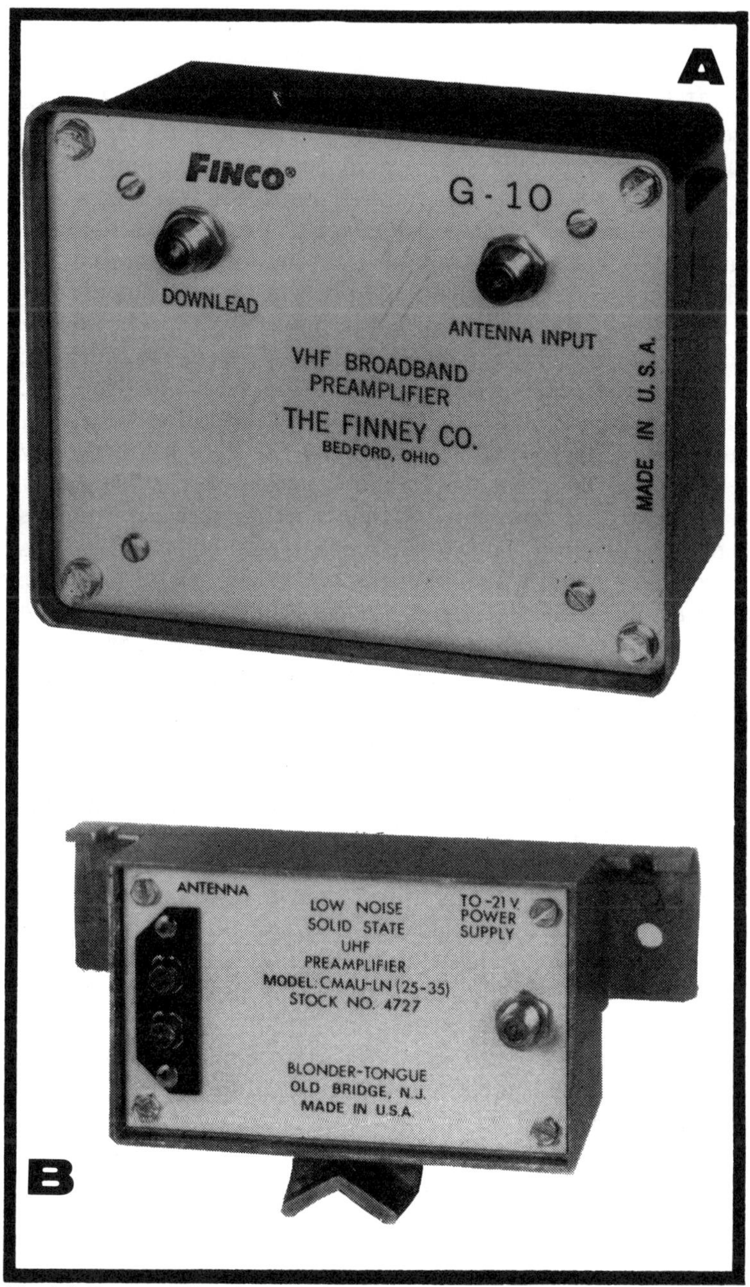

Fig. 2-8. Preamplifiers typical of those available for MATV systems. Note that units are packaged in weather-resistant cast housings for extended outdoor service. A, broadband VHF unit, courtesy The Finney Co. B, low noise UHF preamp, courtesy Blonder-Tongue.

Bandwidth

Here the manufacturer tells you what band(s) of frequencies each model covers. If you are looking for an all-channel preamp, this spec would read one of many ways.

A. 54 to 890 MHz. This implies that there is a single amplifier within the unit tuned to cover all frequencies between the limits given. This type is usually the least expensive and will give excellent service in 50 percent or more of normal MATV jobs. This type of preamp, however, is prone to difficulty where strong signals other than TV can cause overload, such as FM stations, hams, mobile radio, public service, etc. Of potential trouble is the unhappy situation where a moderately strong VHF channel, say Ch. 7, is harmonically related to a UHF channel. Channel 7 picture carrier frequency is 175.25 MHz. If moderately strong it will generate harmonics within the preamp. The 4th harmonic would be 701 MHz. This could ruin Ch. 52, which happens to be from 698 to 704 MHz. Odd harmonics like the 3rd or 5th are generally too weak to be offensive. Even harmonics are the troublesome ones.

B. 54 to 216, 470 to 890 MHz. This spec implies that the unit contains two separate amplifiers, one for VHF, the other for UHF. It could also mean that filters have been included in one broadband amplifier to eliminate any possible interference from frequencies between 216 and 470 MHz. If separate amplifiers are in the box you need not worry about self-generated, harmonically related interference. Two amplifiers also require a band-separating filter from a common input terminal. These filters introduce some loss which adds directly to the N.F. of the preamp thus robbing some sensitivity. Although this is usually small, 0.5 to 1.0 dB, it is worth considering in weak signal areas.

C. 54 to 88, 174 to 216, 470 to 890 MHz. Similar to 'B' above but probably contains three separate amplifiers. Note that the FM band is not covered.

D. 54 to 108, 174 to 216, 470 to 890 MHz. A minor variation on 'C' above to include the FM band.

Gain

Gain is necessary in every amplifier but it is far less important than many other specifications. The function of gain is to raise the antenna signal strength to a sufficiently high level whereby the noise contribution of following amplifiers can be ignored. Gain is also required to overcome the loss of signal in lead-in cable.

Acceptable ranges for gain in a preamp are from 10 to 12 dB minimum, to about 30 dB maximum. Selection depends upon system needs and lead-in cable loss. **Too much gain can be just as bad as too little.** This will be explained later.

Input & Output Capability

This is an important but very misleading specification. Properly used it should describe the minimum signal level needed to provide a picture quality of some specific **TASO** rating. Unfortunately most manufacturers use **input capability** to specify the maximum signal level before preamp overload occurs. This is misleading because a properly designed amplifier will not overload on the input stages but rather on the output stages. Should a particular unit have 3 or 4 dB more gain than minimum spec, it will go into overload at 3 to 4 dB below maximum rated input.

Properly designed amplifiers are limited by the maximum signal-handling capacity of the output stage. At this point the signals are at their highest amplitude and require large, expensive transistors. If limiting occurs in the smaller, less expensive, input or intermediate stages, the amplifier would be said to be poorly designed.

Distortion

There are three types of limiting distortion for MATV amplifiers. For single-channel devices it is either **intermodulation** or **sync compression**. For broadband devices carrying two or more channels, it is **cross modulation between channels**.

Briefly, **intermodulation** is the beat between carriers that falls within the channel being carried. The worst one for single-channel preamps is the beat between the sound carrier and the color subcarrier which will produce a 920 kHz beat, visible in the picture.

The sync portion of a TV signal is the highest amplitude part of that signal. Sync compression or **sync clipping** occurs when the signal level gets high enough to place the synchronizing pulses into the nonlinear portion of the transistor curves. This is usually stated as the **limiting output spec** for a single-channel device since it's the easiest to measure. Output maximums are usually stated at the ½ dB sync compression point.

Cross-modulation is the type of distortion that limits broadband preamps. Here some of the modulation information of one channel appears superimposed on one or more of the other channels going through the amplifier. Typically this distortion appears on the screen as bars wiping back and forth across the screen and has been often called the **windshield wiper** effect. When checking output capability of an amplifier, make certain that the number of channels are also specified along with the cross-mod spec. Usually VHF amplifiers are specified with seven channels and all-channel amplifiers are specified with ten channels. Some manufacturers give maximum signal performance for various numbers of channels, and this makes determination of absolute limits for your area quite easy.

Sometimes, troubles are encountered with single-channel preamps which appear to be operating well within manufacturer's ratings. Single-channel preamps are often selected to aid reception of a weak distant channel with a cut-to-channel antenna. If there is a strong local station just one or two channels away, cross-mod can easily occur. Even two channels away, local signals can still be received at sufficiently high levels to get through the skirts of the preamp tuning and arrive at the output, equal to or even stronger than the desired signal. This produces troubles because the two-channel cross-mod point of any amplifier occurs at approximately 12 dB below the 0.5 dB sync clip point of a single channel. The cure for this is to trap out the unwanted local signal between the antenna and the preamp rather than complain that the preamp doesn't meet specs.

Noise Figure

Noise figure is a preamp's most important specification. As explained in Chapter 1, N.F. is the measure of how much noise any particular amplifier adds to the system. Thus, the higher the N.F., the higher the noise generated by the amplifier. Conversely, the lower the N.F., the better the quality of picture developed from a given signal level. Therefore N.F. becomes the prime measure of product quality for most preamps. Under 6 dB VHF, and under 8 dB UHF are are good N.F.

Preamps under 3 dB N.F. are classified as premium and you will have to pay extra for them. Since N.F. is not an easy measurement to make and it requires special equipment, don't be fooled by claims of premium quality at bargain basement prices! Of course, advances in component technology are providing better quality at lower prices each year. Most manufacturers keep abreast of these improvements and a little comparison of price vs N.F., will usually tell the story. But **beware!**

Of interest is the fact that 3 dB N.F. is the absolute limit for a well matched preamp operating at normal temperatures. The only way to get below this limit is to give up some or all of the input match. This can cause reflections that lead to ghosting if antenna down-leads are not kept short. Also of interest is the fact that atmospheric noise starts to come into play at about this point. For low VHF channels (2 through 6), improvement in N.F. below about 4 dB cannot be seen on the screen due to the atmospheric noise received by the antenna. Such improvements, however, can be seen for the high band VHF and the UHF band.

The remainder of preamp specifications are generally self-explanatory and need no further comment. One word of caution seems in order, however. Most preamps are powered from a remote power supply over the same cable that carries the amplified signals. For the most part, manufacturers employ low voltage ac for this power. Some attempts have been made to reduce the cost of certain models of preamps by using low voltage dc for powering. Under these circumstances, take care to avoid the use of dissimilar metals along the power path to avoid electrolysis. Since this is difficult to assure in all cases, low-voltage powering is not recommended.

MASTER AMPLIFIERS

Like preamps, master amplifiers are available in a wide variety of sizes and functions. Basically the function of the master amplifier is to raise the antenna signals to a level sufficient to serve all the TV sets connected to the MATV system. Master amplifiers are available as single-channel devices as well as broadband. Fig. 2-9 shows some typical amplifiers of the master type. Unlike preamps, master amplifiers are designed to operate at high levels close to their

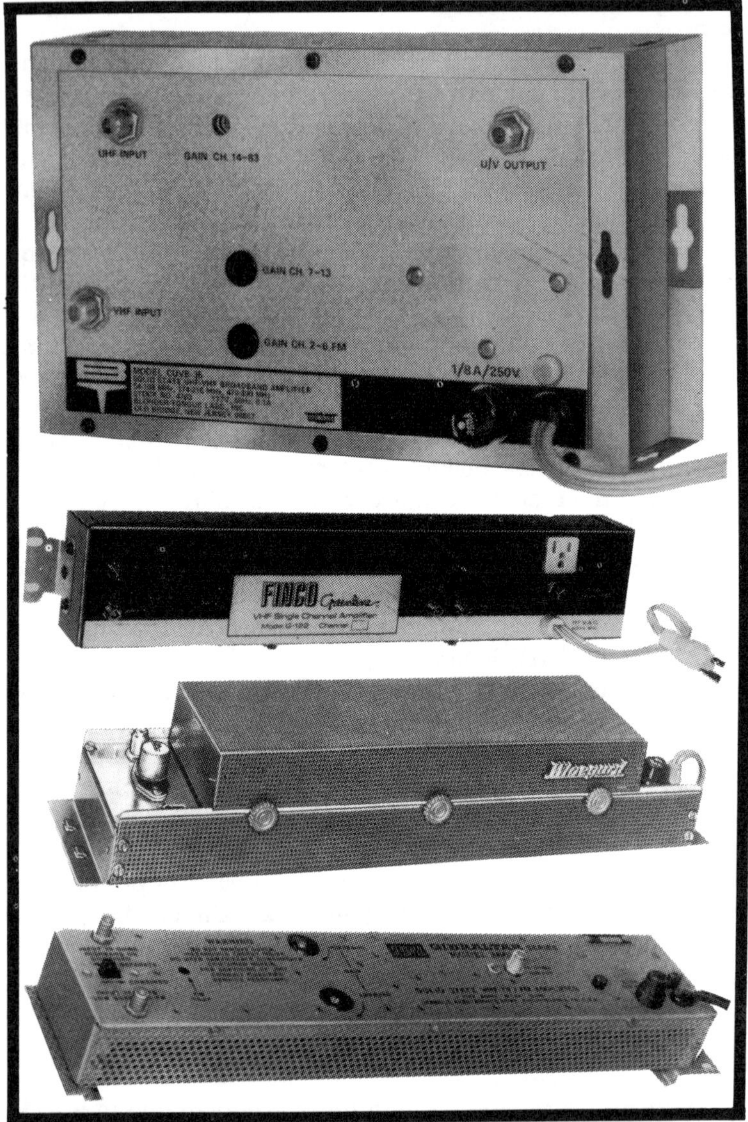

Fig. 2-9. Illustration of the various types of master amplifiers available for MATV service. A. All-channel broadband amplifier with separate gain controls for low VHF, high VHF, and UHF. Model CUVB-35, courtesy Blonder-Tongue.
B. Single-channel VHF strip amplifier with manual gain control. Model G-122, courtesy The Finney Co.
C. Broadband UHF amplifier. Model DA-440, courtesy Winegard.
D. Broadband VHF amplifier with optional separate or combined input terminals. Unit includes tunable FM trap. Model 3661, courtesy Jerrold.

limit of undistorted output. Output capability therefore becomes its most important specification.

Bandwidth

Bandwidth is a spec with all the same implications as discussed for preamps. Selection of single-channel amplifiers requires knowledge about signals being received and level of signals required to serve the system. If antenna signals are strong, steady, and relatively equal in strength and if output requirements are less than 60 dBmV per channel, a broadband amplifier is the best choice. If antenna signals are weak and fluctuating from fading or if levels above 60 dBmV are needed for the system, single-channel "strip" amplifiers should be selected.

Noise Figure

Input signal levels needed to operate the amplifier close to rated output are sufficiently high to preclude concern about noise contribution. The N.F. should not be completely disregarded and figures up to 10 dB VHF and 15 dB UHF are quite acceptable.

Gain

This is obviously an essential requirement in a master amplifier. Each unit should have enough gain to reach maximum permissible output with moderate input signals. A good rule of thumb for gain is that it should be no more than 20 dB less than rated output capability. For example, if a particular amplifier has a rated output of say 51 dBmV, the gain should not be less than 31 dB. For this case it would require an antenna signal level of 20 dBmV to reach full output capability. This level is not unusual in metropolitan areas.

On the average, manufacturers usually supply a gain that is about 10 dB less than rated output and provide a gain control for adjusting operating levels. Gain specs can be somewhat misleading in that manufacturers state the minimum gain with controls set to maximum. Since transistors vary from unit to unit, it is not unusual to find products exceeding their spec by as much as 5 dB. Gain control ranges are also a little misleading in that they mean the amount of reduction from full gain. Thus if a 40 dB gain amplifier has a 10 dB gain control range, you would expect to be able to lower the gain to 30 dB. If, however, this particular unit happens to be a hot one with say 44 dB of actual gain, you would only be able to reduce the gain to 34 dB. This is offered as information only because there are tolerances in every manufacturer's output, and controlling the absolute gain in every product would add considerably to its cost for no justifiable reason.

Output Capability

In master amplifiers, this is the most important specification. This is the spec that must equal or exceed the requirements for the system (as calculated in Chapter 3).

Cross-Mod

In Chapter 1 we were able to turn to an established authority (TASO) for the definition of picture quality vs SNR. Unfortunately there is no similar authority to cite for picture quality vs **cross-mod.** To establish a frame of reference for cross-mod, subjective tests were made where many people were asked to judge picture quality under controlled conditions. After a large number of tests were made, it was determined that for pictures with cross-mod reduced to minus 46 dB below the carrier level, that was the point where the majority of people said they could no longer see the interference. During the test, however, everybody reported seeing the familiar windshield wiper effect of cross-mod when the picture faded to a blank screen. Further tests showed that with a blank screen everybody reported that they no longer saw cross-mod when it was reduced to minus 51 dB below picture carrier level.

These are two very worthwhile points to remember. **Blank screen tests** are sometimes found in proof of performance specs for large jobs. My question is: **who watches blank screens?** This gives rise to the popular use of minus 46 dB cross-mod as the limiting distortion for amplifier output. Some manufacturers use minus 40 dB cross-mod as their limiting spec under the assumption that the amount of interference at this level is so small that nobody complains.

Cross-mod also encompasses the number of channels involved. With ten channels, cross-mod will occur at a much lower output level than with three channels. Fig. 2-10 is a graph showing the permissible variations in operating levels for an amplifier that is spec'd at 51 dBmV per channel for seven channels. Many things can be learned from this graph. If you go along with the minus 40 dB cross-mod rating, this amplifier should be rated for 54 dBmV per channel output. If you must design the system to pass a blank screen test, you would limit per channel output to 48 dBmV with seven channels. At recommended minus 46 dB cross-mod and twelve channels, output maximum would be 48.25 dBmV. If only three channels were being carried then outputs could be raised to 54.5 dBmV without exceeding the cross-mod limit.

This graph shows the wide variations possible in specifying the operating limits of only one amplifier. Any point on the graph may be selected to describe the amplifier's capability and **all are truthful**! Since no standard method has evolved within the MATV industry, it is important to understand this, especially when reading and comparing published claims. You may also use this graph to determine actual operating limits for various products with the number

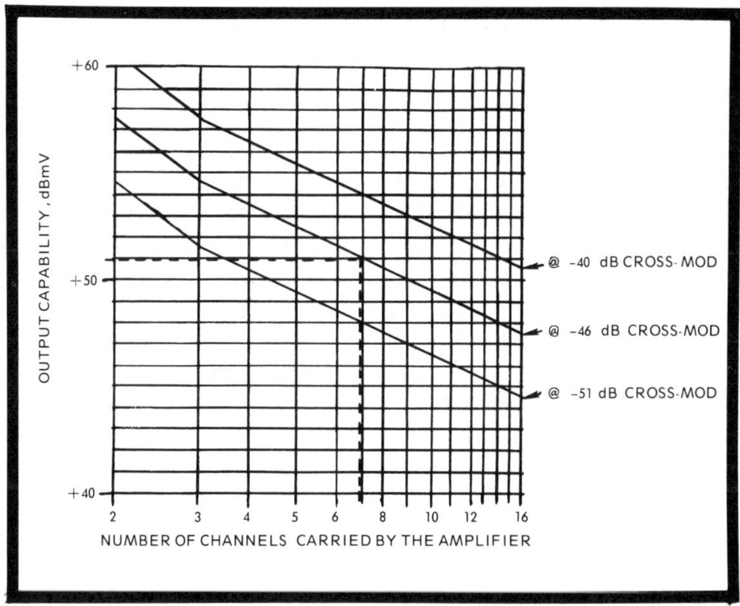

Fig. 2-10. Graph compares output capability of amplifier rated at 51 dBmV with 7 channels at -46 dB cross-mod with permitted variations for other than 7 channels and other specifications for cross-mod.

of channels in your area. Just substitute different numbers onto the output capability scale to match the rated specs for any amplifier in question.

Output capability for a single-channel strip amp is not limited by cross-mod because only one channel is being amplified. As in the discussion on preamps, single-channel output is limited by intermod beats or sync compression. Sync compression is easily understood in that it's the level where the sync tips reach into the amplifier's nonlinear region and get clipped off.

Some very old TV sets would lose vertical hold if the sync was compressed much more than 0.5 dB. Modern TV sets will hold vertical sync with most of the sync pulse missing. Now we in MATV face a real dilemma. One might think that the 1 dB spec is better because it permits higher operating levels for the amplifier. True enough for black and white television. Not true with color. You'll read **why** in a moment.

Intermod Distortion in Color TV

With color TV it is now necessary to concern ourselves with **intermod distortion**. This is the resulting beat between the color subcarrier and the sound carrier that appears in the picture as a herringbone pattern. In transistorized amplifiers, this 920 kHz beat becomes visible at 2 to 4 dB below the 0.5 dB sync compression point. This throws suspicion on the entire method of using sync compression as a measure of output limitation. Sync compression is still used as a spec because it's much easier to measure than intermod. This also accounts for the fact that most manufacturers recommend operating their products at 2 to 4 dB below maximum output.

Intermod is not limited to interference within the channel being carried. Beats between the picture carrier and sound carrier also produce interference beats that fall into the adjacent channels immediately above **and** below. This beat can be ignored unless the system in question happens to be carrying adjacent channels. Here again the only way out is to reduce the operating levels by 2 to 4 dB below max.

AGC

Automatic gain control is typically found on strip amps. Some strips are offered with manual control and these serve many useful functions. Automatic gain control is employed to maintain a constant level of operation into a system where the input signal is not steady. This is typical of signals received from more than about 30 miles from the transmitter. The further out the receiving antenna is from the transmitter, the wider the range of signal fading encountered. Rarely in MATV work does this effect exceed about 20 to 25 dB.

Automatic gain controls function in a strip amp in a very straightforward manner. Typically, a tuned circuit in the amplifier output samples a portion of the pix carrier. This sample is detected, filtered and amplified to some dc level equivalent to the pix carrier strength. This dc voltage is fed back to some of the earlier stages in the amplifier to control their gain. A front panel control lets you adjust the output level by varying the agc feedback voltage.

Amplifier gain is always specified as **maximum**. The agc range tells you how much the amplifier can reduce its own gain automatically. Consider an amplifier with the following specs: Output capability 72 dBmV, gain 55 dB, agc range 20 dB. It would require an input signal level of 17 dBmV for the amplifier to achieve full output. Any signal fading under these conditions would cause the output to decrease proportionately because the amplifier is using all its specified gain. If the input level were 27 dBmV and the output adjusted to 72 dBmV, then the amplifier would be working at only 45 dB gain. This would permit the input signal to rise or fall as much as 10 dB from nominal level while the output remained essentially constant.

Signal fading should be taken literally. In distant reception situations there is some signal level which represents the average condition. Signal strength from the antenna will fluctuate from this level but mostly in a **downward** direction. Take a tough situation where signals vary some 20 dB. If you watched the level over a long period of time you would find an interesting thing. Increases in level **above** the average would be limited to approximately 3 to 5 dB. Decreases in level would be as much as 15 dB below the average. This unequal variation is common to the vast majority of situations. It gives us a clue as to the best method of operation under such conditions.

In the example cited earlier, the input signal level at 27 dBmV placed the amplifier in the middle of its operating range. It the input signal faded 15 dB, the output would remain constant for the first 10 dB of the fade, then fall off for the last 5 dB. A better average signal input would be 32 dBmV thus allowing the signal to vary plus 5 dB and minus 15 dB while remaining within the full capability of the amplifier's agc.

The spec agc "stiffness" or tolerance is the manufacturer's way of saying that the output level will vary slightly over the full range of input variation.

CABLE

Coaxial cable (coax) is the transmission line used to carry signals from place to place in an MATV system. Coax is available in many sizes and types of construction and each has its place in MATV work. Graph 6 in the appendix illustrates the loss characteristics of all popular sizes of coax used in MATV work. Fig. 2-11 shows the structure of a foil coax.

Loss

Loss per 100 ft is undoubtedly the most important specification for coax. Note that the loss for any given cable increases with frequency. Also note that the loss at any given frequency is proportional to cable size. Since cable cost is related to size, this gives the system designer a clear mandate to select coax carefully. Generally speaking, the smallest cable that will not exceed the maximum losses permitted in a

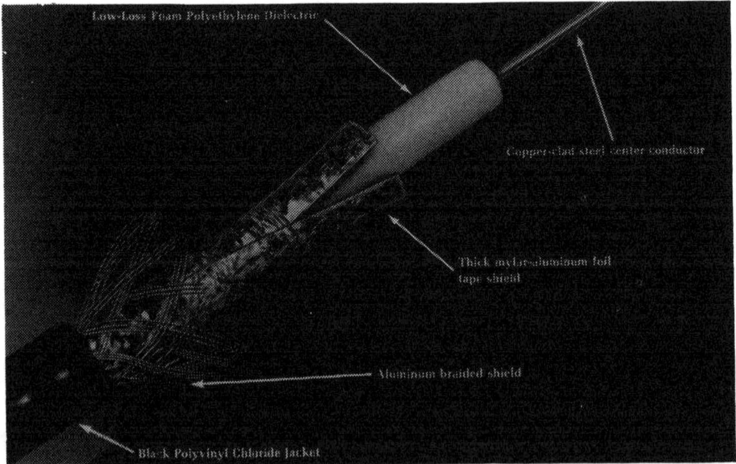

Fig. 2-11. Cutaway showing coax cable construction, model CAC-6, CAC-11, courtesy Jerrold Electronics.

design, is the proper selection. For smaller jobs of about 25 outlets, RG-59 size foam or solid dielectric coax is best suited. For larger jobs, RG-6 is a good all around choice. In any job where UHF is distributed on-channel, RG-6 foam dielectric coax is almost mandatory.

Match

Loss is not the only property of coax. **Match** is definitely a measurement of its quality. Match is its ability to carry signal without producing objectionable reflections from discontinuities in the coax's contruction. Only **sweep tested** cables are recommended for MATV work. Match specifications of 26 dB at VHF and 20 dB at UHF are considered **minimum**. Special physical construction is also important in determining which cable is best suited for any particular job.

Center Conductor, Dielectric, & Shield

Usually this is solid copper wire of the proper gauge for each size cable. Lately, cables with copper clad steel centers have been offered and have proven their worth. The solderless connectors used in MATV use the center conductor to mate directly with chassis fittings. Copper clad withstands the rigors of frequent connections and won't buckle if forced. Nonbuckling is important; buckling shorts the braid to the center conductor. The rf performance is unimpaired since in a good coax, the rf energy travels along the copper covering (skin effect).

Insulating material between center conductor and shield is offered in two major forms, solid polyethylene or foamed polyethylene (poly). Foamed poly produces the lowest loss per foot and is most widely used. Foam poly has one drawback. If exposed to constant moisture such as being buried directly in the ground, it will absorb moisture which will collect in the foamed cells and increase cable loss drastically. Therefore foamed poly cable should not be used underground unless properly protected.

The cable shield performs many functions. It is the second wire of the two wires necessary to carry any electrical signal. In coax, rf is carried on the inside of the shield, because of skin effect. It also shields these signals from all outside interferences. Last but not least, it gives the cable mechanical strength. Of late, many cables have appeared which use an aluminized plastic tape wrapped around the dielectric as a shield. This tape provides a substantial improvement in rf performance, lower loss and much higher shielding than woven braid. The tape, however, offers little mechanical strength. Two approaches to this problem are (1) drain wires or (2) partial braid second shielding. **Drain wires** are usually four small alumi-weld wires helically wrapped along the outside of the tape. This offers some mechanical strength and is adequate for work where the cable is supported in trays or in open construction such as false ceilings. It will **not** stand the rigors of pulling through conduit or support itself in vertical runs. Braid-covered cables will stand up under these conditions and is better suited for general MATV work although it is somewhat more expensive.

Jacketing

The finished cable is protected with a plastic covering of vinyl or polyvinylcloride (PVC). Vinyl is a long lasting material but is frowned upon in MATV work because it will support combustion. The PVC will not support combustion but will deteriorate if exposed to sunlight. The best material is termed **noncontaminating** and is made of PVC with lamp-black added to stop the ultra-violet rays of sunlight. This results in a cable that is black in color, but this is considered a small penalty to pay for quality merchandise.

Underground wiring has become more and more popular as MATV systems are built for garden court apartments and mobile home parks. Direct burial of cable is the least expensive approach to this type construction. For this type of work there is a full range of seamless aluminum shielded cables available. Cables that are PVC-jacketed and filled with a moisture barrier flooding compound can be expected to give 20 or more years of direct burial service. Since long distances are usually involved, these cables are offered in large diameter, low loss sizes. Popular sizes are 0.412, 0.500, and 0.750 in. which represents the outside diameter of the aluminum shield. Mating splice connectors and adapter fittings to standard 'F' fittings are available for each cable size.

SPLITTERS AND COUPLERS

The many legs required to supply signal to all branches of a system are created by splitting the output of the MATV head end amplifier. Determining which type splitter to use in what combination is discussed in Chapter 3. **Splitters** are those devices that split signals equally to all outputs, and there are usually two or four outputs per splitter. **Couplers** are the units which split the signal unequally to two outputs. These are available in a variety of "sizes" from 8 to 16 dB.

All of these devices are constructed with coils and transformers that have a limited bandwidth of operation. Most modern devices cover the full TV spectrum of 54 to 890 MHz. Older devices, still in use, were limited to 54 to 216 MHz and care must be taken when adding channels to older systems, especially UHF channels.

Characteristic of all splitters and couplers is the circuitry that makes them all a form of directional coupler. Unique to this type of circuit is the large amount of **isolation** between output terminals. Fig. 2-12 illustrates this action. In the two-way splitter, loss from input to either output is typically 3.5 dB. The outputs are isolated from each other by 15 dB or more due to the directional circuitry. The advantage is that anything that should happen on one output leg is well isolated from the other. This unit can be used as a mixer, such as when combining adjacent-channel strip amps where isolation between outputs prevents detuning of output filters.

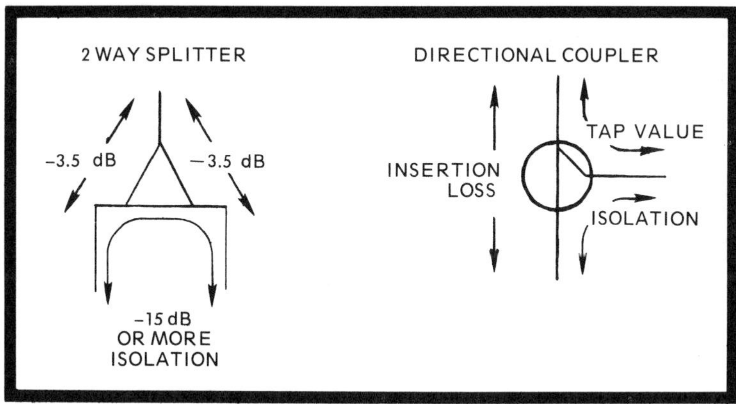

Fig. 2-12. Illustrating the meaning of insertion loss and isolation and loss in splitters and directional couplers compared with MATV standard symbols.

The **directional couplers** act the same way except that the splitting losses are not equal and descriptive names are given to the terminals. As with the splitter, the directional coupler can also be used as a mixer. In this application the two signals to be mixed go into the terminals marked **out** and **tap**. Combined signals appear on the **in** terminal. Couplers are generally used to mix signals of unequal level where the tap loss helps to balance their levels. A typical application would be mixing a closed-circuit camera with the signal from an antenna. The coupler value is choosen to attenuate the camera signal and the large isolation prevents any unwanted radiation of the camera signal back through the antenna.

TAPS AND TERMINAL DEVICES

Taps are the devices used to provide signals for the connection of TV sets to the distribution system. Taps do just as their name implies, they tap the line and bring a portion of the signal out to the TV. Fig. 2-13 is a simplified circuit diagram of a typical tap. All tap-off devices should contain these basic ingredients in one form or another. Multiple-outlet taps contain either separate tap circuits for each outlet or a splitter circuit after one tap circuit, all within a single enclosure.

The illustration shows a resistive network bridging the feeder line. The amount of signal delivered to the tap is dependent upon resistor value. This resistor isolates the tap from the line by some specified amount which determines the tap's **isolation value**. This network may also be a capacitor, a

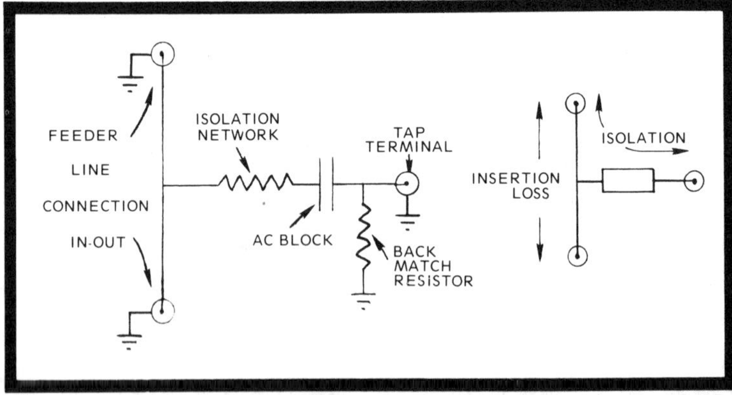

Fig. 2-13. Schematic and block diagram of typical MATV tap showing essential components and illustration of common terms, insertion and isolation.

transformer, a directional coupler circuit, or a combination of elements. Whatever, the function is the same, to sample a fraction of the signal on the line and make it available to the tap.

The capacitor is used to provide ac isolation between the tap line and the feeder line. This isolation protects the system and all other connections to the system from any accidental power that may get fed back into the tap from the external connection. Power isolation may be provided in other ways, such as a transformer, but is strongly recommended for all taps. A few taps are available without ac isolation. These are usually the real bargain basement items where pennies were pinched to achieve low pricing. Beware—but also know that in many cases this will cause no problem since something else provides the isolation function, maybe the matching transformer or the TV set itself. Since that's not under your control and the tap is, the tap is the right place to insist on proper power isolation.

The resistors to ground represent a load to the tap and perform the function called **back match**. Back match is needed to absorb any reflections that might occur on the external connection. TV sets are not necessarily well matched. Even well-matched sets are only a good match at the channel to which they are tuned when turned on. Other signals are free to bounce back and forth on the external line with no place to go. Without this back match circuit, ghosting can occur although pix smearing or changes in color performance are more often seen. Since this circuit robs energy and costs money to build in, it too is often left out in price line taps.

Tap Specs

Insertion loss is the amount of loss that the tap adds to the line in which it is installed. **Isolation loss** is the amount that signals will be reduced between the high level feeder line and the tap output.

Flush-mount taps are used for installation in electrical outlet boxes with concealed wiring. **Surface-mounted taps** are those used with exposed wiring such as along baseboards or in enclosed areas such as hung ceilings, etc. This is also the type available with multiple outlets.

Pressure taps are a special mechanical arrangement whereby the tap samples the signal within the cable by probing the center conductor through a hole drilled into the feeder cable. The tap is held in place with a block clamped to the cable. This type of tap is often used in external wiring jobs. **Multitap** is the term used to describe CATV type taps used on poles or in underground pedestals for systems such as mobile home parks or housing complexes. This type tap is generally available with two, three, or four outlets and is completely protected for outdoor applications. Fig. 2-14 shows these different mechanical features.

Through Match

This is an important part of tap specifications. Any device such as a tap that is cut into a line has the possibility of introducing a discontinuity. Discontinuities in the line produce reflections that add up to trouble. In MATV work, taps are often spaced at regular intervals, such as every 10 ft between floors. Small reflections to the signals passing through the taps can add together to produce serious difficulties. Periodic reflections such as this can produce pix smearing, loss of color, modification of expected signal level, and in long runs, possibly the complete loss of one or more channels.

Taps with good through match prevent this from happening. Good through match specs are 18 dB or more for VHF and 15 dB or more for UHF. This does not preclude the effective use of unmatched taps such as pressure taps. Pressure taps introduce a discontinuity that cannot be compensated for within the tap design. These taps can be used if you avoid periodic spacing. Physically locating these taps at random intervals means that the reflections will be random and will not add together to produce visible degradation.

Wall Terminals

Wall terminals are not taps but convenient devices for terminating the end of a cable. The name is misleading in that the device provides a mechanical termination for the cable but not an electrical termination. It's merely a convenient device for easy connection of TV sets to the line that ends in the room.

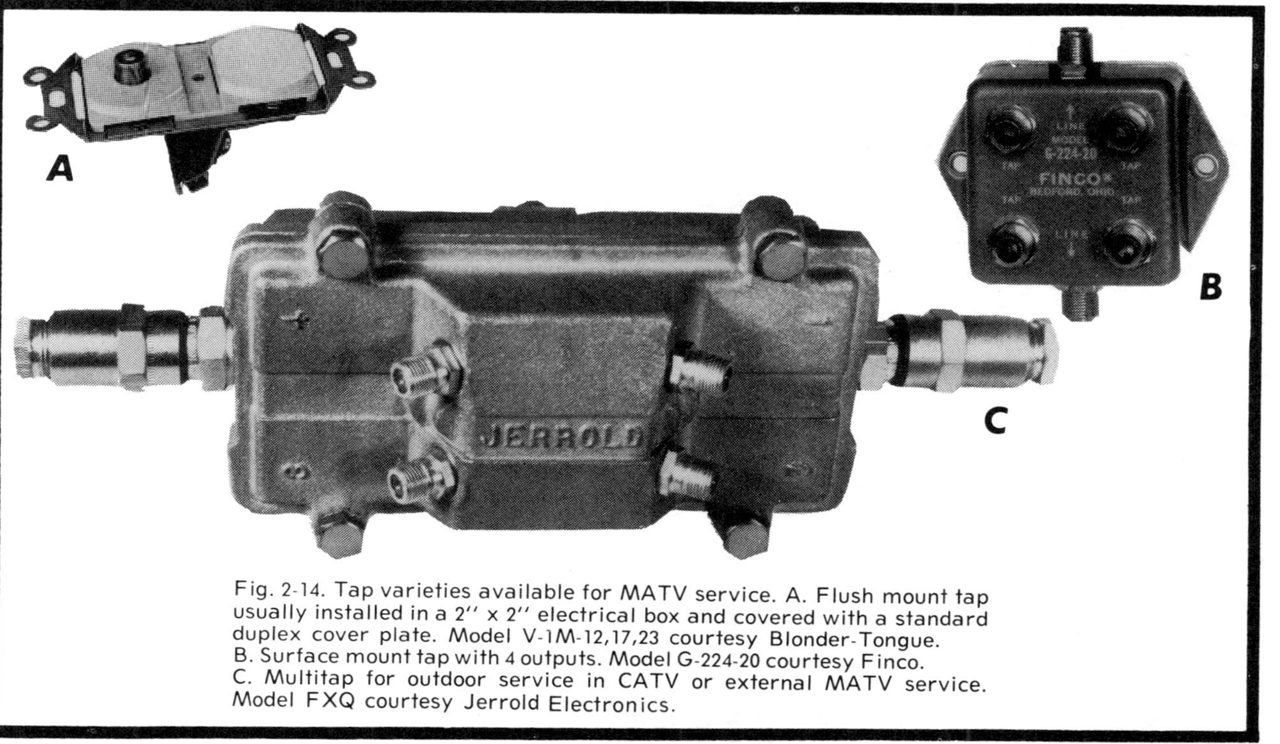

Fig. 2-14. Tap varieties available for MATV service. A. Flush mount tap usually installed in a 2'' x 2'' electrical box and covered with a standard duplex cover plate. Model V-1M-12,17,23 courtesy Blonder-Tongue. B. Surface mount tap with 4 outputs. Model G-224-20 courtesy Finco. C. Multitap for outdoor service in CATV or external MATV service. Model FXQ courtesy Jerrold Electronics.

MATCHING TRANSFORMERS

The matching transformer at the back of the set is often referred to as a terminal device. Again the termination is mechanical, not electrical. Transformers perform the important function of matching the 75-ohm unbalanced coax to the 300-ohm balanced input of the TV set. Matching transformers are also used to match 300-ohm antennas to 75-ohm MATV equipment.

Matching transformers come in two types, regular two-winding transformers and the more complicated four-winding balun. Two-winding transformers are ac isolated from input to output but are poorly balanced. Baluns are not isolated unless capacitors are included, but they are highly balanced.

Balance is very desirable in MATV work where direct pickup is a problem. This prevents any external braid currents from the coax acting like a long wire antenna from getting into the TV. (See "overcoming direct pickup" in Chapter 5 for more details.) Balance of greater than 30 dB can provide real protection from this difficult system problem. Fig. 2-15 illustrates two typical transformers used in MATV. The T-28 by Winegard is obviously packaged for use outdoors with 300-ohm antennas. Model 4540 by Blonder-Tongue is a combination UHF-VHF splitter and matching transformer for use at the back of the TV set.

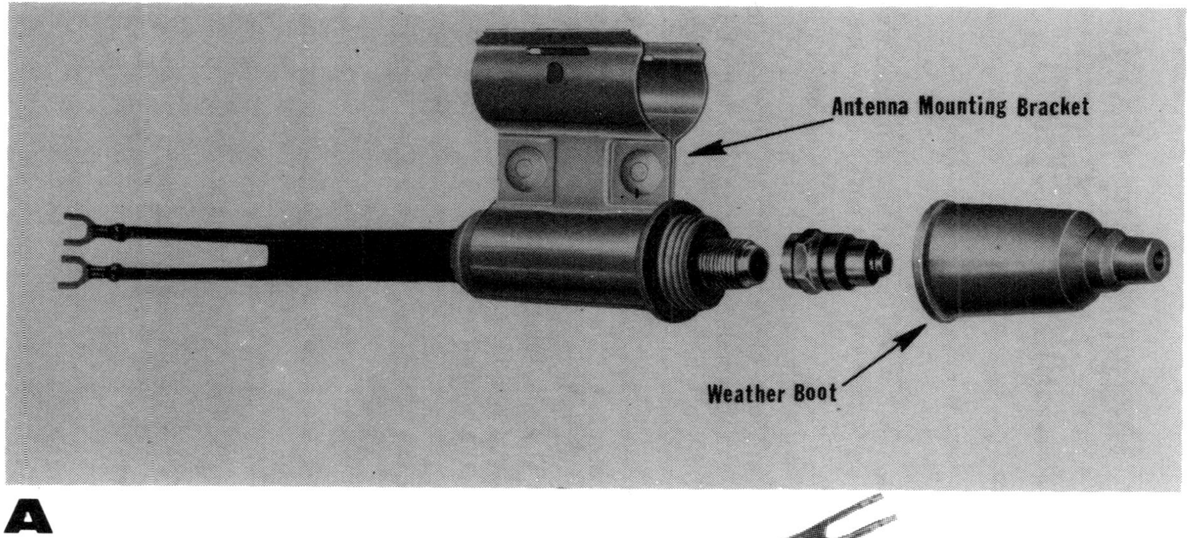

Antenna Mounting Bracket

Weather Boot

A

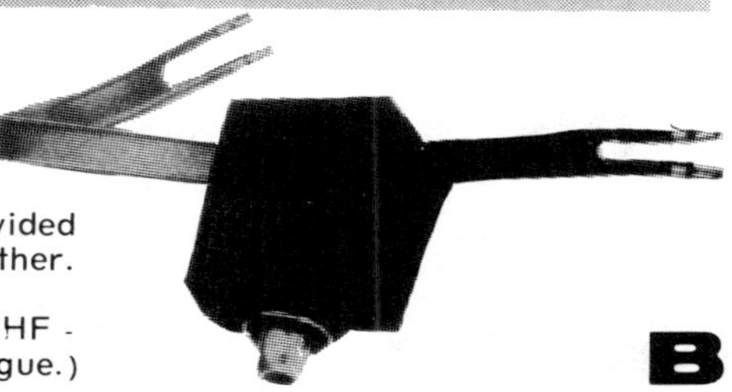

Fig. 2-15. Matching transformers.
A. Model T-28 is equipped for mounting on an antenna boom and provided with a weather boot to protect the cable connection from the weather. (Courtesy Wiregard.)
B. Model 4540 is a 75-ohm to 300-ohm matching transformer with a UHF - VHF band separator built into one package. (Courtesy Blonder-Tongue.)

3 Designing the MATV Distribution System

I never cease to be amazed when confronting a prospective customer who wants an MATV system but hasn't the faintest idea of what it will do. This is really not surprising when you stop and think that this is really why you're in business. You bring to this customer your unique ability to provide the service he wants. It is small wonder then that you will have to do a great deal of detail work before you can begin to design the system.

Each type of system has its special considerations; however, they all have most of the same common problems. It's up to you to dig out the information which will lead you to a workable design. A lot of forethought can save many headaches! Not only must you think out the technical details—you must also quote a competitive price.

ANTENNA LOCATION

You must resolve several questions relative to location for antennas: How near is the head end equipment? Is conduit available? Must the roof be penetrated for cable lead-ins? Does the owner want the antennas hidden from public view? What type of mounting will suit the building structure? Is the roof area open to building tenants, especially children, etc.?

EQUIPMENT LOCATION

Is there power available? Do the conduits for TV stub up in the right places? Is the area secure from building tenants? Will you have easy access to the equipment for servicing? And, in new construction, when will this area be available to you?

BUILDING WIRING

Here are the tough ones. How will your cables run—in conduit, in false ceilings, in crawl spaces, or in floor duct? In buildings with horizontal runs, how do you get from floor to floor vertically? Most important is an accurate determination of cable footage from the head end to the end of every cable. What is the spacing between outlets?

A set of plans for the building, drawn to scale, is the most desirable starting point. TV outlet locations will generally be called out. If not, you must meet with your customer and choose them. Don't depend on plans to give you conduit-routing information. This is usually left to the discretion of the electrical contractor and in many cases, the straight line is not the correct distance. Sometimes conduits must be run via indirect routes to avoid interference with building structure or other services such as heating ducts, etc.

All of these mechanical questions must be answered with equal importance to true MATV questions such as: How many channels are to be carried? How much signal strength is available on the roof? What are the directions to the stations? Is there a direct pickup problem at the location? Some of these questions the customer can answer, but most of them will require advice from you in helping the customer to understand what he can expect from his MATV system.

There are a number of special considerations which you should explore with the customer. In many cases you will be calling his attention to things he didn't consider. For example:

In apartment buildings is there a need for a door surveillance camera? Must FM be distributed to each outlet?

In hospitals, is remote control of the TV necessary—what about private listening with pillow speakers?

In hotels and motels, is there a background music system in each room or can you provide this service on one or more unused TV channels?

In schools, is it necessary to distribute all the entertainment channels to each classroom or can you provide entertainment channels selectively, using only one cable channel and leaving all the rest for educational purposes?

We will explore many of these questions in detail. We will also discuss some additional ones through the following chapters. The importance of knowing what the system is expected to do before starting the design will become evident as we proceed with general design considerations.

CALCULATING THE SYSTEM LAYOUT

In many respects, each MATV distribution system is unique. The exact number of outlets is known, the cable footage is known, the routing of cable is known, and the expected results are well defined. But if given all this information, or if the information is ferreted out, the precise equipment needed to meet all engineering objectives can be calculated, if the designer starts at the right place. In MATV, the right place is the most remote TV set to be served by the system. The most efficient design will develop if you work **backward** from that last outlet, toward the head end.

Calculations for an Apartment Building

Consider the layout illustrated in Fig. 3-1. For the sake of simplicity, we assume a building with equal length runs in all riser cables. This is fairly typical for a medium- or high-rise building. We will calculate this system for VHF-only distribution. Channel 13 is the highest frequency channel to be carried.

Since the objective is to deliver good signal quality to the last set on the line, we must choose a minimum level. From the discussion on SNR in Chapter 1, we learned that 1000 uV or 0 dBmV was needed to assure snow-free operation. This level becomes the starting point for the calculation.

Leaving the back of the set, the first thing encountered is usually a matching transformer and a short run of coax to the wall outlet. The transformer and say 10 ft of cable have some loss. Average transformer losses are usually about 0.5 dB. It is now necessary to select a type of coax cable because its loss must be included. Graph 6 in the appendix gives the loss characteristics of many common types of cable used in MATV. In VHF systems, it is common to use a good grade of foam dielectric RG-59. This usually gives the lowest cost installation. Looking up the loss of foam RG-59 we find Ch. 13 at approximately 4.0 dB per 100 ft or 0.4 dB for 10 ft. This says that the signal level out of the wall tap must be 0.9 dB stronger than the 1000 uV objective, to overcome the losses considered thus far.

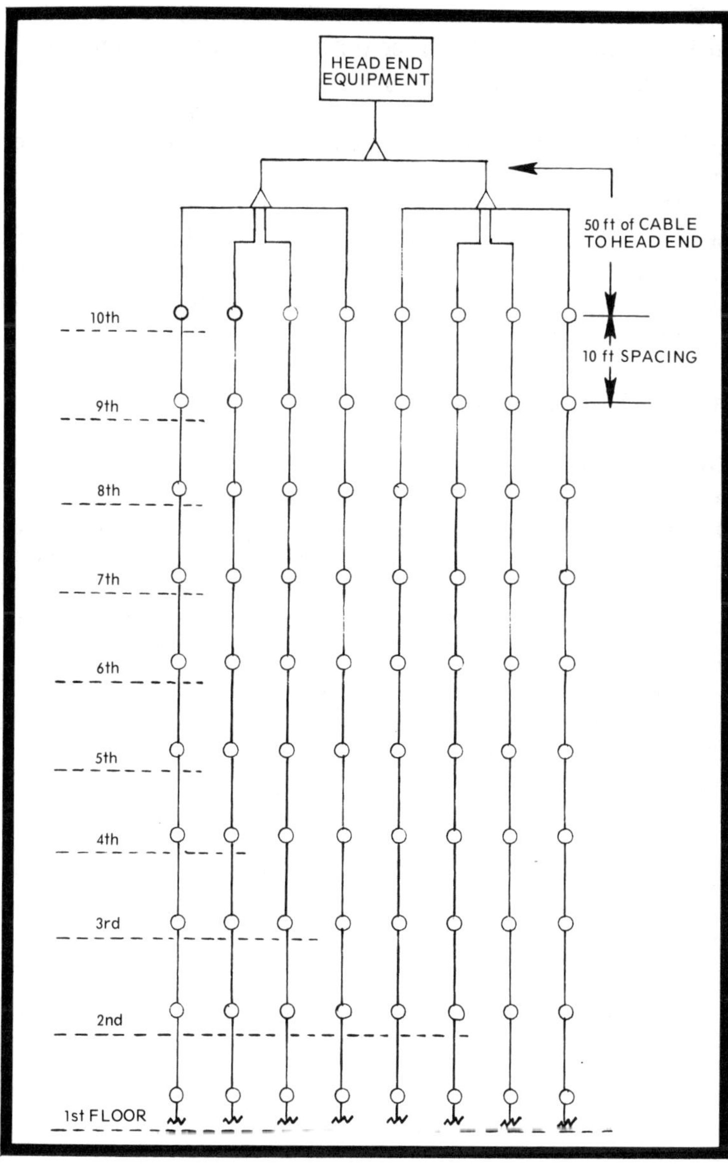

Fig. 3-1. MATV distribution layout for an 80-outlet system in a 10 story building. Calculations of this system are shown in Fig. 3-3.

Next it is necessary to select a wall tap. Here, three things govern.

1. The tap must be made to fit into a 2 x 4 in. wall box if that's what's provided for the TV wiring, as is usual.

2. Do you connect the set cable to the tap with a threaded 'F' fitting or provide a customer-removable push-on connection, G fitting, or Motorola type jack?
3. What are the electrical characteristics of the tap? Is it available in enough different values to give essentially the same signal level at all outlets?

Fig. 3-2 is a table of tap specifications typical of available products. Note that at Ch. 13 the values of isolation are 23 dB, 17 dB, and 12 dB. It is logical to choose the smallest isolation available for the last tap. This will require the least amount of signal from the head end amp.

We can now determine how much signal is required in the riser at the last tap by simply adding the isolation loss of this tap to the signal required out of the tap. This would be 0.9 dBmV plus 12 dB isolation loss equals 12.9 dBmV signal level required in the riser. Next to consider is the 10 ft of cable between the last tap on the first floor and the next to last tap on the second floor. Again we add 0.4 dB to the signal required to overcome the loss of the cable. Thus 12.9 dBmV plus 0.4 dB cable loss equals 13.3 dBmV at the output of the next to last tap. Consulting Fig. 3-2 we find that model 3 taps have an insertion loss of 0.9 dB. Since the signal from the head end must pass through this tap, we must include it in our calculation, 13.3 dBmV plus 0.9 dB insertion loss equals 14.2 dBmV required signal into the next to last tap to assure not less than 0 dBmV to the last TV set on the system.

You must continue this process of accounting for all losses that signals encounter on their way from head end to outlet so that you can determine how much total signal is required from the head end amp to do the job. Fig. 3-3 illustrates the completed calculation. Only one riser line is shown for simplicity.

			CHANNELS			
MODEL			2	6	7	13
1	Isolation	dB	24	24	23	23
	Insertion	dB	0.3	0.3	0.3	0.3
2	Isolation	dB	19	18	17	17
	Insertion	dB	0.6	0.7	0.7	0.7
3	Isolation	dB	15	14	13	12
	Insertion	dB	0.8	0.8	0.9	0.9

Fig. 3-2. Specifications for MATV system taps used in the calculations of Fig. 3-3.

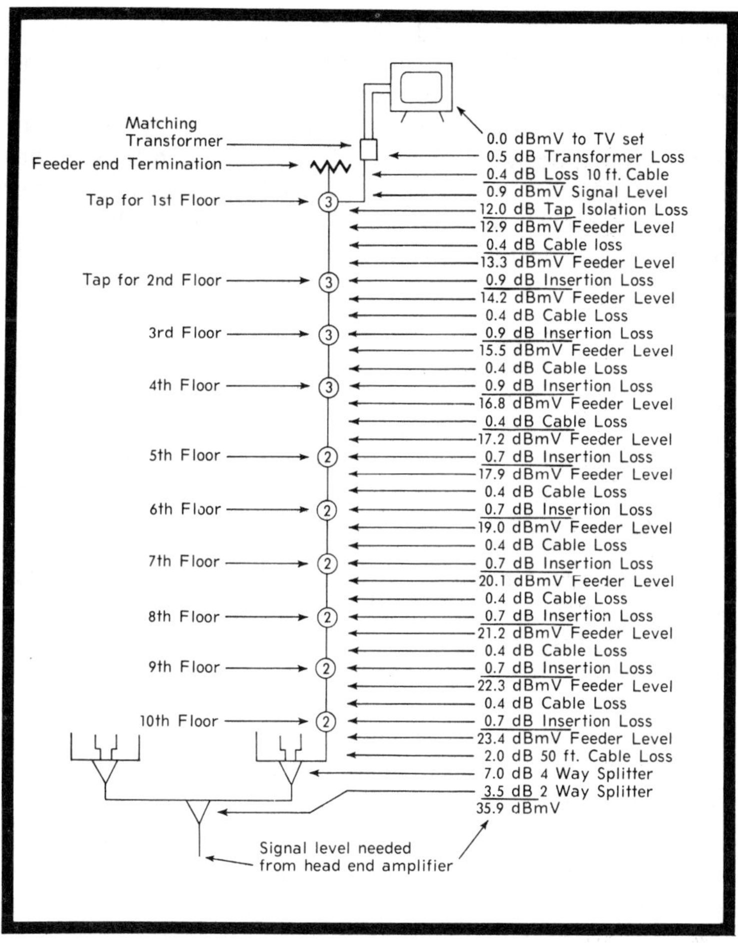

Fig. 3-3. Calculation of one feeder line of the system layout of Fig. 3-1. Calculation of only one feeder is necessary since all other feeders are exactly the same.

Also note that the layout is upside down so that corresponding parts of the calculation are adjacent to each layout symbol.

By following the tabulation you can see that model 3 taps are used for the 1st, 2nd, 3rd, and 4th floors. At the 5th floor the signal level required at the output of the tap is shown separately. Note that this level is larger than the isolation value of a model 2 tap. This is the clue to when you should change to the next highest value of tap. The reasons for this are two-fold: First, with a model 2 tap, the signal delivered to the TV set on the 5th floor will be 0.9 dBmV or **more** than the

0 dBmV minimum required. Second, the insertion loss of a model 2 tap is smaller than a model 3, thus the signal required from the head end will be smaller.

This method of calculating a system layout gives you the most efficient design possible with the cable and taps chosen for the job. This will permit you to use the smallest amplifier possible to meet your engineering objectives. It also means that you have designed for the lowest cost in system equipment.

Note that the signal level into the tap on the 10th floor is 23.4 dBmV. This is not quite large enough to permit a change from model 2 to model 1 taps. If a model 1 tap were used, the signal levels delivered to the set on the 10th floor would be minus 0.9 dBmV. This is shy of the engineering objective. If this layout had an 11th floor it would be proper to use the model 1 tap there.

The total requirement for signal must include the 50 ft run of cable from the head end to the start of the riser. It must also include the losses of splitters required to create the eight building risers. It can be seen that the signal level to each riser is the same as to the one calculated. Since the requirements for all eight risers are identical, there is no need to duplicate the calculation. But you must remember to include costs for eight runs in your quotation!

This method of calculating a system layout gives you a lot of information. When you order equipment for the job, you know just how many of each model tap to purchase. In this case, you need 32 model 3 taps and 48 model 2 taps. Your calculations should be shown on the blueprint so that the installation crews will know what value tap to put on each floor. When you activate the system, you should test outlet levels with a meter. If there is reasonable agreement with the calculations, you can be sure the job was installed correctly. If not, the measurements recorded compared with your "should be" data can help you locate the trouble.

The design of the head end itself is covered in Chapter 4. Since the head end cannot be tackled until the distribution system requirements are known, we will continue with layout calculations. The next example will illustrate a different type of building and will use a different wiring method.

Calculations for an Elementary Grade School Building

Consider the partial floor plan of the modern type elementary school building shown in Fig. 3-4. An investigation of the building finds it to be a one-story structure. All building wiring is routed above the false ceiling of the central corridors. Each of the 20 classrooms is to be equipped with two outlets for TV in the front of the room. Conduit has been stubbed into the false ceiling area adjacent to each room from standard electrical outlet boxes. A secure location for the head end equipment is spotted in a small storage closet adjacent to the gym.

The most logical approach to wiring this type of structure is a feeder cable down each hallway. At every location where

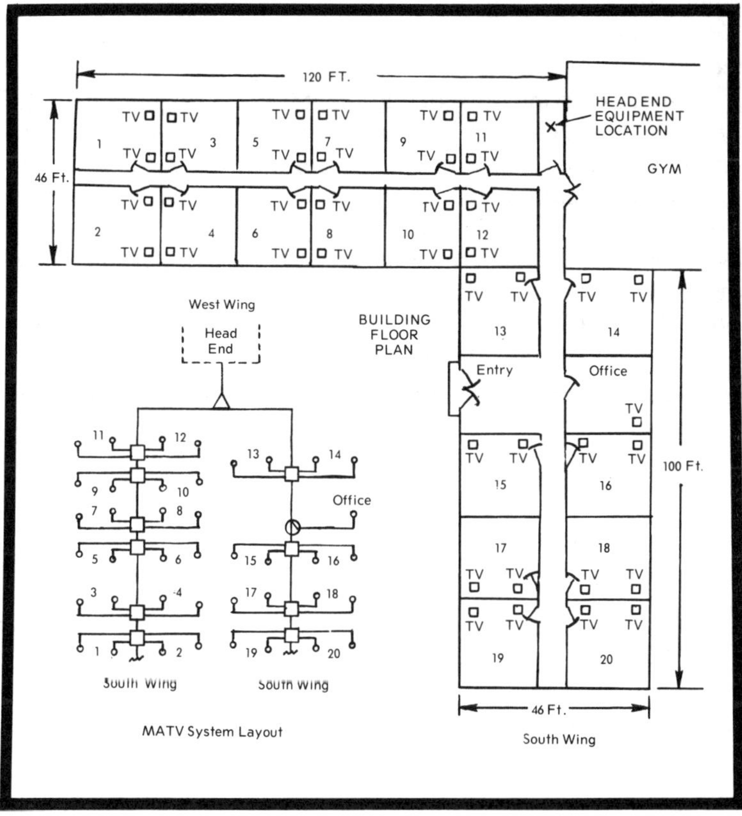

Fig. 3-4. Partial floor plan for an elementary school showing location of TV outlets. MATV system layout shown at lower left. See text for method of calculation.

outlet conduits appear, you should locate a tap. A cable from the tap runs through each conduit to a wall terminal or feed-through plate behind each TV. Since schools and other public buildings are built with public funds, there is undoubtedly a set of specifications covering the TV requirements. We will assume for this example that the specs call for 2000 microvolts at each outlet. Assume also that VHF only is to be distributed.

By looking at the floor plan again we observe two more bits of information. First we notice that the TV outlets are grouped in fours and eights. This calls to mind the four-outlet multitaps offered by many suppliers.

Second, we note that the dimensions of the two school wings are not identical and the number of outlets is different for each wing. To assure a proper layout it will be necessary to calculate individually the requirements of both wings. This calculation is approached in the same way as the previous example.

Before starting the calculation, it is necessary to select the type of cable to be used. Also required is the performance data on the multitaps. Since this is a small system with VHF-only distribution, foam insulated RG-59 can be used as before. Fig. 3-5 is a table of tap specs for four-way multitaps. Note that the isolation values are the same as in the first example. The insertion losses, however, are considerable higher. This is typical of multitaps because for all practical purposes there are four individual taps within each unit.

			CHANNELS			
Model			2	6	7	13
A	Isolation	dB	24	24	23	23
	Insertion	dB	1.0	1.0	1.0	1.0
B	Isolation	dB	19	18	17	17
	Insertion	dB	2.0	2.0	2.0	2.0
C	Isolation	dB	15	14	13	12
	Insertion	dB	3.0	3.0	3.2	3.3

Fig. 3-5. Specifications for MATV system taps used in the calculation of system for Fig. 3-4.

The following is the calculation of signal level requirements for each feeder. Look them over carefully and follow each against the wiring layout in Fig. 3-4.

West Wing	
+6.0 dBmV	to last TV outlet
—1.0 dB	Approx. 25 ft. cable tap to outlet
+7.0 dBmV	At tap output
—12.0 dB	'C' tap isolation
+19.0 dBmV	Input to last tap
—3.3 dB	'C' tap insertion, Rm 3&4
+22.3 dBmV	input level
—1.6 dB	40 ft. feeder cable
—2.0 dB	'B' tap insertion Rm 5&6
+25.9 dBmV	Input Level
—2.0 dB	'B' tap insertion Rm 7&8
+27.9 dBmV	Input Level
—1.6 dB	40' feeder cable
—2.0 dB	'B tap insertion Rm 9&10
+31.5 dBmV	Input level
—1.0 dB	'A' tap insertion Rm 11&12
+32.5 dBmV	Input level
—2.0 dB	Approx. 50' feeder cable
+34.5 dBmV	required feeder level

South Wing	
+6.0 dBmV	to last TV outlet
—1.0 dB	Approx. 25' cable tap to outlet
+7.0 dBmV	at tap output
—12.0 dB	'C' tap isolation
+19.0 dBmV	Input to last tap
—3.3 dB	'C' tap insertion, Rm 17&18
+22.3 dBmV	input level
—1.6 dB	40' feeder cable
—2.0 dB	'b' tap insertion Rm 15&16
+25.9 dBmV	Input Level
—0.5 dB	Directional coupler for office
+26.4 dBmV	input level
—1.6 dB	40' feeder cable
—2.0 dB	Tap insertion, Rm 13&14
+30.0 dBmV	Input Level
—2.0 dB	Approx. 50' feeder cable
+32.0 dBmV	Required feeder level

You will note that each leg requires a different signal level to meet specs. The difference, however, is small and there is no reason to use other than a two-way equal splitter at the master amp output to feed signals to each wing. This adds 3.5 dB to the larger requirement for 38 dBmV total. Signals for the office on the south wing are supplied by a directional coupler. This is one of a number of practical approaches to odd, individual outlet, requirements. Most manufacturers offer single, dual, and four-way taps for applications such as this example and a proper isolation value single tap could have been selected.

VHF and UHF Distributed on One System

The MATV systems need not be limited to distribution of VHF channels (2-13) alone. There are surprisingly few locations in the U.S. where at least one UHF station cannot be received. Modern TV receivers are all equipped with tuners that cover Channels 2 through 83. Most manufacturers now offer all-channel equipment and cable for MATV systems. The following example takes advantage of these facts and illustrates how a large job can be done including on-channel distribution of UHF. It is recommended that the system design be limited to the highest channel to be carried. It can be seen by studying the loss characteristics of cable, that there is considerable difference between the loss of any one cable at Ch. 14 vs the loss at Ch. 83. If all systems were designed for Ch. 83 operation, much of the cost of such designs would be wasted if only two or three of the low number channels are carried. Take care to include in the design any known or suspected new UHF channels of higher frequency than currently on the air. A call to the national or regional offices of the FCC should be of help in this regard.

Let us consider the sample layout for a 12-story apartment building with 102 outlets on nine risers, as shown in Fig. 3-6. Note that three risers contain two outlets per floor and five risers call for one outlet for each floor. Six extra outlets are required on a separate riser for lower floor commercial areas. In this example we will assume on-channel distribution of all channels with Ch. 48 being the highest UHF channel. For this job we will select the lowest loss cable that is compatible with available wall taps. We'll use foil shielded RG-6 foam. This is the largest cable that can be used with available wall taps. At 680 MHz, the frequency of Ch. 48, this type cable has a loss of approx. 6.0 dB per 100 ft. Specifications for the taps are given in the table of Fig. 3-7.

Calculations for 12-Story Apartment Building

Again, the starting point for the calculation is the last tap on each riser. Since there are three different types of risers, three separate calculations must be made. Assume no unusual circumstances and design for 0 dBmV minimum outlet signal level.

NOTES:

Iso, is isolation
Ins, is insertion
H.E. is Head End

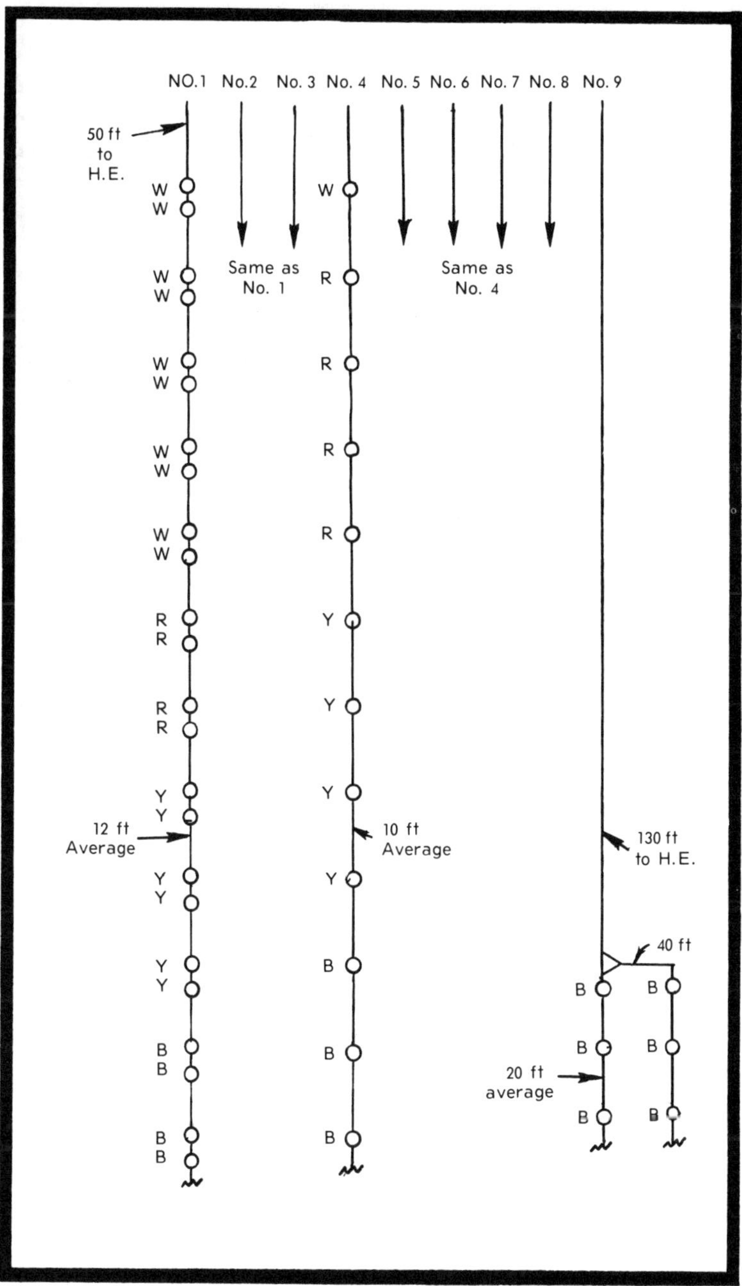

Fig. 3-6. MATV distribution system design problem: In 12 story high rise building with 138 outlets, distribute all signals on-channel including UHF. See text.

MODEL			CHANNELS 2	13	14	48	83
White	Isolation	dB	30.5	30	29	27	25.5
	Insertion	dB	0.3	0.3	0.3	0.3	0.3
Red	Isolation	dB	24.5	24	23.5	23	22
	Insertion	dB	0.6	0.6	0.6	0.6	0.7
Yellow	Isolation	dB	21	19	18	17	16
	Insertion	dB	0.8	0.8	0.8	0.9	0.9
Blue	Isolation	dB	17	13.5	12.5	12	11.5
	Insertion	dB	1.2	1.2	1.2	1.3	1.3

Fig. 3-7. Specifications for MATV system taps used in the calculation of system for Fig. 3-6.

Tap abbrev. (See Fig. 3-7)
W — white
R — red
Y — yellow
B — blue

For dual outlet risers

0.0 dBmV	At last TV outlet
12.0 dB	B tap No. 1 iso. 1st Flr.
1.3 dB	B tap No. 2 ins. 1st Flr.
+13.3 dBmV	Riser level 1st Flr.
0.7 dB	12 ft. cable loss
2.6 dB	2 Blue taps ins.
+16.6 dBmV	Riser level 2nd Flr.
0.7 dB	12 ft cable loss
1.8 dB	2 Yellow taps
+19.1 dBmV	Riser level 3rd Flr.
2.5 dB	12 ft cable & 2 Y taps
+21.6 dBmV	Riser level 4th Flr.
2.5 dB	12 ft cable & 2 Y taps
+24.1 dBmV	Riser level 5th Flr.
1.9 dB	12 ft cable & 2 R taps
+26.0 dBmV	Riser level 6th Flr.
1.9 dB	12 ft cable & 2 R taps

+27.9 dBmV	Riser level 7th Flr.
1.3 dB	12 ft cable & 2 W taps
+29.2 dBmV	Riser level 8th Flr.
1.3 dB	12 ft cable & 2 W taps
+30.5 dBMV	Riser level 9th Flr.
1.3 dB	12 ft cable & 2 W taps
+31.8 dBmV	Riser level 10 Flr.
1.3 dB	12 ft cable & 2 W taps
+33.1 dBmV	Riser level 11th Flr.
1.3 dB	12 ft cable & 2 W taps
+34.4 dBmV	Riser level 12th Flr.
3.0 dB	50 ft cable to H.E.
+37.4 dBmV	Riser level required

For single outlet risers

0.0 dBmV	At last TV outlet
12.0 dB	B tap iso. 1st Flr.
+12.0 dBmV	Riser level 1st Flr.
0.6 dB	10 ft cable loss
1.3 dB	B tap iso. 2nd Flr.
+13.9 dBmV	Riser level 2nd Flr.
1.9 dB	10 ft cable & B tap
+15.8 dBmV	Riser level 3rd Flr.
1.5 dB	10 ft cable & Y tap
+17.3 dBmV	Riser level 4th Flr.
1.5 dB	10 ft cable & Y tap
+18.8 dBmV	Riser level 5th Flr.
1.5 dB	10 ft cable & Y tap
+20.3 dBmV	Riser level 6th Flr.
1.5 dB	10 ft cable & Y tap

+21.8 dBmV	Riser level 7th Flr.
1.2 dB	10 ft cable & R tap
+23.0 dBmV	Riser level 8th Flr.
1.2 dB	10 ft cable & R tap
+24.2 dBmV	Riser level 9th Flr.
1.2 dB	10 ft cable & R tap
+25.4 dBmV	Riser level 10 Flr.
1.2 dB	10 ft cable & R tap
+26.6 dBmV	Riser level 11th Flr.
0.9 dB	10 ft cable & W tap
+27.5 dBmV	Riser level 12th Flr.
3.0 dB	50 ft cable to H.E.
+30.5 dBmV	Riser level required

For commercial outlet riser

0.0 dBmV	At last TV outlet
12.0 dB	Blue tap isolation
+12.0 dBmV	Riser level 1st Flr.
1.2 dB	20 ft cable loss
1.3 dB	Blue tap insertion
+14.5 dBmV	Riser level 2nd Flr.
2.5 dB	20 ft cable & B tap
+17.0 dBmV	Riser level 3rd Flr.
2.4 dB	40 ft lateral cable run
4.0 dB	2-way splitter loss
+23.4 dBmV	Riser level into splitter
7.7 dB	130 ft cable to H.E.
+31.1 dBmV	Riser level required

Determining Total Head End Output

Now that the minimum requirements **for each riser** have been calculated, it is necessary to combine them to determine the total head end output needed. This should be accomplished in the most efficient way. This is done by grouping together risers of equal signal requirement and combining them with two- or four-way splitters as needed. Fig. 3-8 is the recommended arrangement for this example. Note that combining four of the single-outlet risers calls for a level nearly equal to level of the three dual-outlet risers. The signal levels shown in dBmV in parentheses are minimum requirements based on the calculations. Numbers shown without parentheses are the actual levels based on meeting the toughest requirement. This arrangement of splitters meets the calculated requirement with a maximum discrepancy of 1.1 dB which indicates a very efficient design.

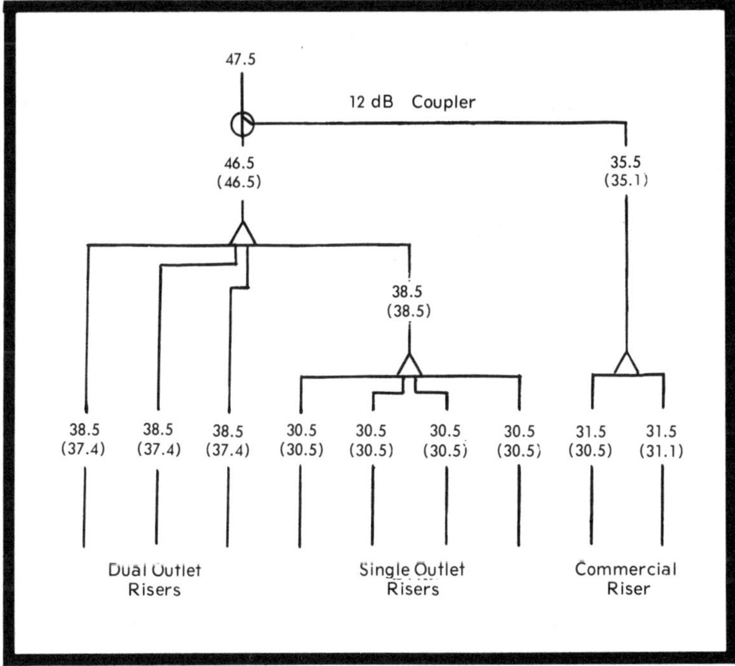

Fig. 3-8. Most efficient design of head end splitter arrangement for the system of Fig. 3-6. Signal levels shown in parentheses are **minimum** levels required to meet layout calculations. Levels not in parentheses are those delivered from the head end to all feeders, based on the worst case requirement.

Checking High- vs Low-Frequency Losses

It has been stated that designing a system at the highest frequency of operation is all that is needed. Since the losses of cable, tap insertion, etc., are generally lower at lower frequencies, this should be correct. Let's check to see if this is true. We might find that too much signal is being delivered and that overload of some TV sets might occur.

To make this check, simply add up all the losses from the H.E. to the outlet being checked. Subtract this total loss from the head end signal and that's what is delivered to the TV. In the example below, we will check Ch. 2 to the last outlet on each of the two major types of risers.

Dual outlet riser

—1.0 dB Coupler
—7.0 dB 4-way Splitter
—3.0 dB 10 W taps
—2.4 dB 4 R taps
—4.8 dB 6 Y taps
—3.6 dB 3 B taps
—2.9 dB 180 ft cable, Ch. 2
—17.0 dB B tap isolation

—41.7 dB Total loss
+47.5 dBmV H.E. output

+5.8 dBmV to last TV outlet, Ch. 2

Single outlet riser

—1.0 dB Coupler
—14.0 dB 2, 4-way Splitters
—0.3 dB 1 W tap
—2.4 dB 4 R taps
—3.2 dB 4 Y taps
—2.4 dB 2 B taps
—2.9 dB 180 ft cable, Ch. 2
—17.0 dB B tap isolation

−44.2 dB Total loss
+47.5 dBmV H.E. output

+3.1 dBmV to last TV outlet, Ch. 2

This check shows us a couple of things. First, the assumption that all would be well at lower frequencies is true. Second, it shows that equal signal level for all channels out at the **head end** is correct and that overloading of TV sets will **not** occur. This is because the manufacturer has graded the isolation value of each tap to be more at low frequencies than at the high frequencies. **Graded isolation** acts as an equalizer to compensate for the fact that cable loss is much lower at these low frequencies.

The method of designing a distribution system has been adequately illustrated in the preceding examples. It will be necessary for you to apply these principles to each new job and probably no two of them will be alike. From here on experience is your best teacher.

Getting More Coverage Without Increasing Signal

There is a limit to the amount of undistorted signal that can be delivered by head end equipment. When designing all-channel systems this limit can be reached rather quickly, especially in high-rise buildings over 20 floors. The following is a little used but noteworthy technique that lets you design larger systems with no increase in signal level from the head end. It's a way to get more coverage out of a given amount of signal.

The technique is illustrated in Fig. 3-9. The proof of the technique can be seen in the following discussion. For the sake of brevity, let's assume that 26 dBmV of signal level was all that could be supplied to the start of a riser. Consider on-channel distribution of highest Ch. 48 to double outlets as in the previous example.

The single riser illustrates a portion of the previous example taken where the calculated signal level was 26 dBmV. The split riser technique shows the same level into a two-way splitter. One leg of the splitter can feed eight outlets at slightly higher, 0.4 dB, levels than before. The other leg feeds an untapped bypass cable which picks up the continuation of taps for three additional floors. Again minimum

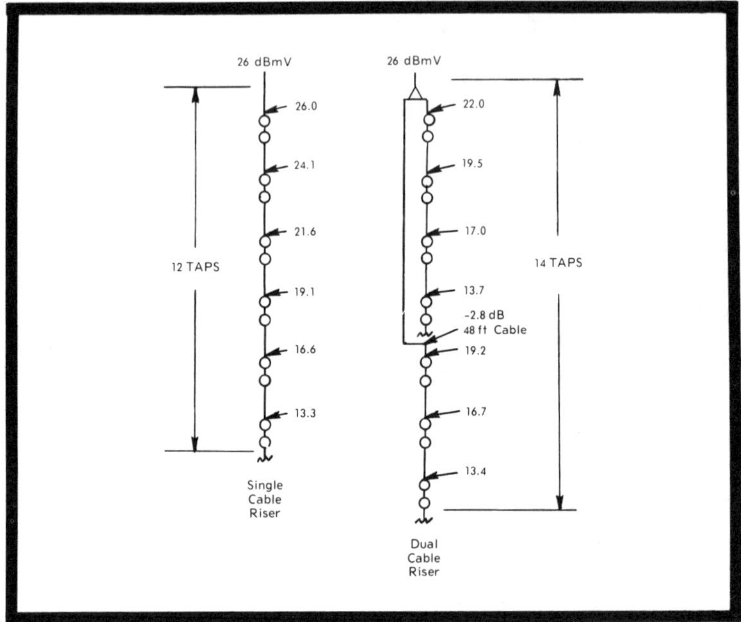

Fig. 3-9. Illustration of how a splitter can be used to increase the amount of coverage from a given amount of signal at the input.

signal levels are exceeded by 0.1 dB. The result is a 15 percent increase, 12 vs 14, in the number of taps that can be serviced from a given riser signal input level. If the starting level were higher and the bypass cable were one of the lower loss cables like RG-11 foam, the increased coverage could approach 40 percent.

CAMPUS TYPE DISTRIBUTION SYSTEMS

Building distribution systems are not the only types needing MATV. There is often the need to design a system for a complex of individual buildings such as a college campus, garden court apartment, or a trailer park. This type of system design is approached differently and is very much akin to CATV systems design.

In large systems, the capabilities of even the highest output head ends are often exceeded by sheer distance. To cope with these long distances requires reamplification of signals along the way; CATV type amplifiers are used for this purpose. Certain MATV broadband amplifiers, if properly operated, may also be adapted to this application.

The starting point for designing such systems is usually the head end. Signals are then divided into as many legs as needed to start the system coverage. Taps are placed along each leg as required. If a line runs low on signal strength before reaching the end, a line reamplifier (reamp) is inserted at an appropriate place. Fig. 3-10 is a system layout for a 290-pad trailer park. Construction is underground direct burial with taps and reamps in pedestals.

The head end is located as shown with four equal-level outlets at 57 dBmV each. The cable is buried along the back lot right of way with multitaps located to serve up to four trailer pads.

The design procedure to be used in this type of layout is to calculate **forward** along each leg. Taps are located at physically convenient points to serve the maximum number of outlets. The isolation value of each tap is selected to deliver a level of 6 dBmV to each outlet depending on signal availability in the feeder line at each point. Taps used in this type of system are usually of the multiple outlet variety, with from two to four outlets available. The 6 dBmV at each tap assures sufficient signal strength into the TV set even after encountering the cable loss between tap and TV.

Advantage is taken of the highest output single-channel strip amps available. The design for the head end is discussed in Chapter 4. The signal strength available to each of the four legs is designed to be 57 dBmV. This is a comfortable level which includes 11 dB for splitter, mixing, and power insertion losses and a few decibels of output derating from maximum capability. The high level of operation will provide the maximum system coverage and minimize the number of reamps needed to cover the entire system. Note that only four **line extension amps** are needed in this design. Note also that only one leg is sufficiently long to require cascading of line amps.

It is wise to hold to a minimum the number of cascaded amps, due to accumulated distortion. Suffice it to say that for the types of systems encountered in MATV work, three amplifiers are usually enough. If system requirements get much beyond this point it will be necessary to go to larger CATV equipment not covered in this basic discussion.

The equipment recommended for this type of system is drawn from the CATV industry. In reality this layout can be looked upon as a miniature CATV system. All equipment used

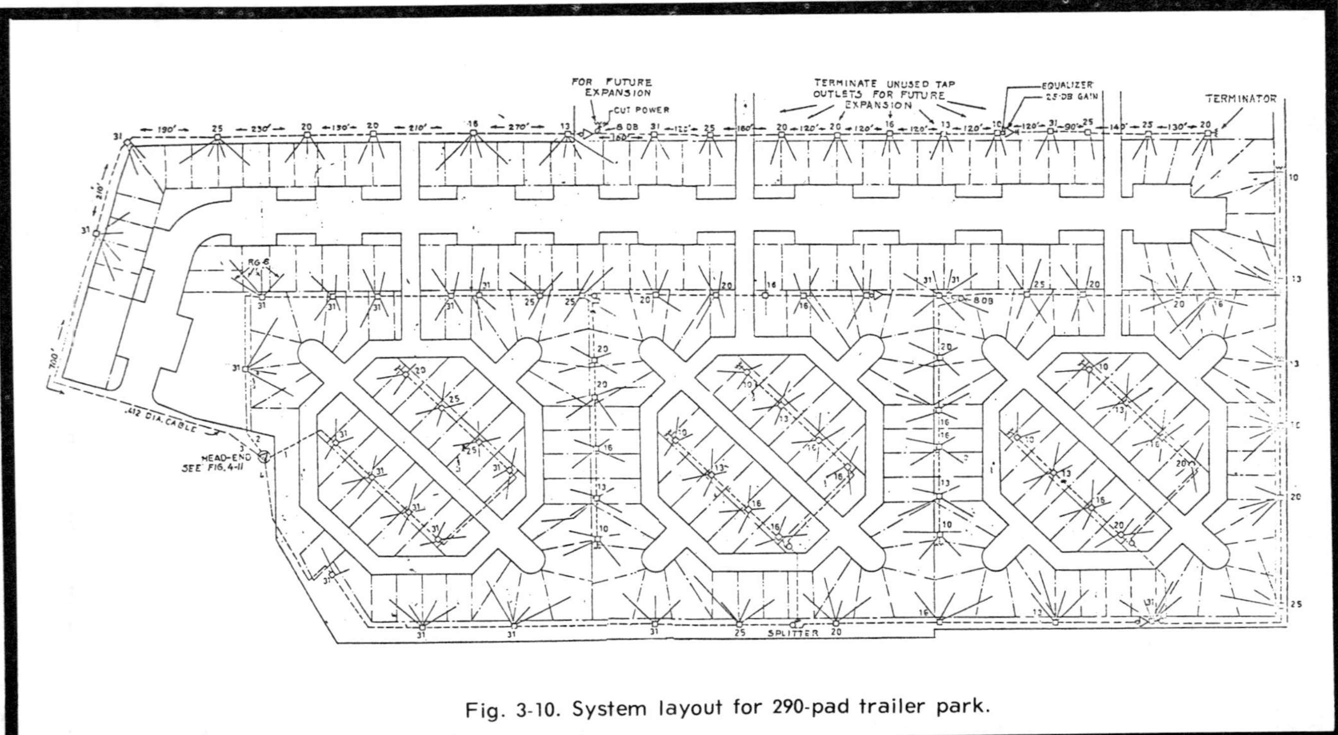

Fig. 3-10. System layout for 290-pad trailer park.

is exposed to the elements and therefore deserves CATV quality. The cable will most likely be buried and must be protected against deterioration. Seamless aluminum tube shielded cable is recommended in all outdoor applications. Polyvinylcloride jacketing with moisture barrier flooding compound is needed for direct burial applications. Standard 0.412 diameter cable is selected for this system and the loss at

Tap Model	Isolation to each outlet, dB	Insertion loss to feeder line, dB
31	31	0.5
25	25	0.6
20	20	0.8
16	16	1.1
13	13	2.0
10	10	3.5

Fig. 3-11. Specifications for CATV type 4-outlet multitap.

Gain	25 dB, min.
Output capability	45 dBmV, 12 channels, -57 dB cross-mod
Noise figure	16 dB, max.
Slope	0 to 8 dB adjustable
Power	22 to 30V AC @ 0.250A via coax.

Fig. 3-12. Specifications for CATV type feeder-line-extender amplifier.

Ch. 13 is specified at 1.62 dB per ft. Fig. 3-11 is a table of tap specs typical of the type used. Fig. 3-12 shows the specs for the line extender amplifier used, again typical of equipment available.

The system calculation is shown in Fig. 3-13. Only the one long feeder with two line amplifiers is shown. All other feeders are calculated similarly.

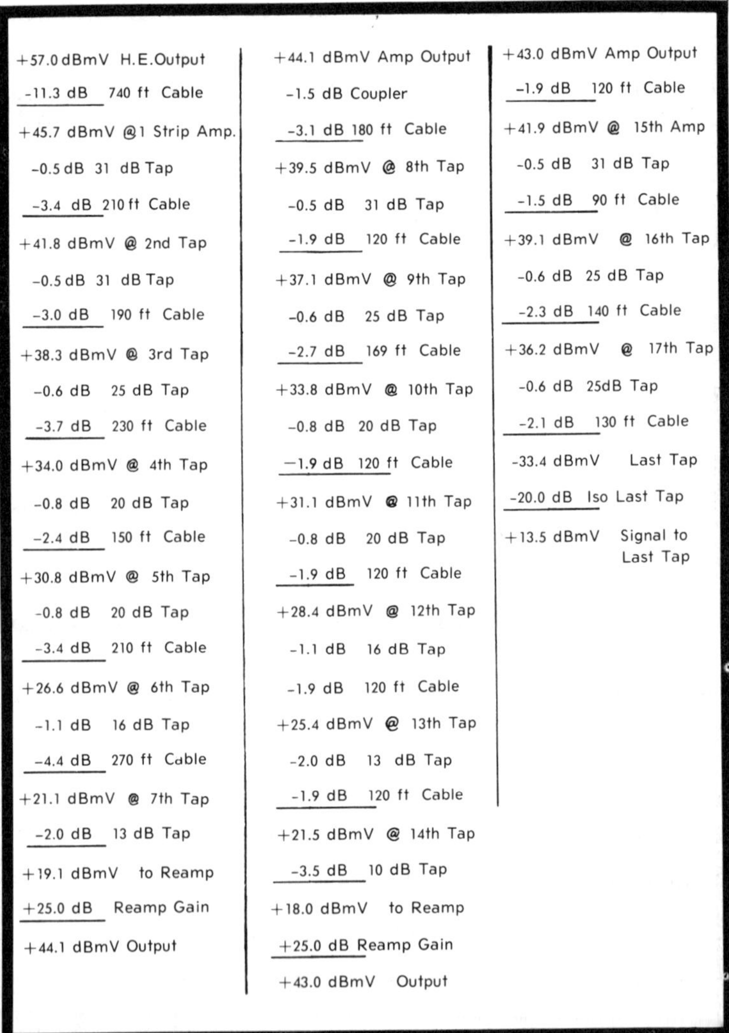

+57.0 dBmV H.E.Output
-11.3 dB 740 ft Cable
+45.7 dBmV @1 Strip Amp.
-0.5 dB 31 dB Tap
-3.4 dB 210 ft Cable
+41.8 dBmV @ 2nd Tap
-0.5 dB 31 dB Tap
-3.0 dB 190 ft Cable
+38.3 dBmV @ 3rd Tap
-0.6 dB 25 dB Tap
-3.7 dB 230 ft Cable
+34.0 dBmV @ 4th Tap
-0.8 dB 20 dB Tap
-2.4 dB 150 ft Cable
+30.8 dBmV @ 5th Tap
-0.8 dB 20 dB Tap
-3.4 dB 210 ft Cable
+26.6 dBmV @ 6th Tap
-1.1 dB 16 dB Tap
-4.4 dB 270 ft Cable
+21.1 dBmV @ 7th Tap
-2.0 dB 13 dB Tap
+19.1 dBmV to Reamp
+25.0 dB Reamp Gain
+44.1 dBmV Output

+44.1 dBmV Amp Output
-1.5 dB Coupler
-3.1 dB 180 ft Cable
+39.5 dBmV @ 8th Tap
-0.5 dB 31 dB Tap
-1.9 dB 120 ft Cable
+37.1 dBmV @ 9th Tap
-0.6 dB 25 dB Tap
-2.7 dB 169 ft Cable
+33.8 dBmV @ 10th Tap
-0.8 dB 20 dB Tap
−1.9 dB 120 ft Cable
+31.1 dBmV @ 11th Tap
-0.8 dB 20 dB Tap
-1.9 dB 120 ft Cable
+28.4 dBmV @ 12th Tap
-1.1 dB 16 dB Tap
-1.9 dB 120 ft Cable
+25.4 dBmV @ 13th Tap
-2.0 dB 13 dB Tap
-1.9 dB 120 ft Cable
+21.5 dBmV @ 14th Tap
-3.5 dB 10 dB Tap
+18.0 dBmV to Reamp
+25.0 dB Reamp Gain
+43.0 dBmV Output

+43.0 dBmV Amp Output
-1.9 dB 120 ft Cable
+41.9 dBmV @ 15th Amp
-0.5 dB 31 dB Tap
-1.5 dB 90 ft Cable
+39.1 dBmV @ 16th Tap
-0.6 dB 25 dB Tap
-2.3 dB 140 ft Cable
+36.2 dBmV @ 17th Tap
-0.6 dB 25dB Tap
-2.1 dB 130 ft Cable
-33.4 dBmV Last Tap
-20.0 dB Iso Last Tap
+13.5 dBmV Signal to Last Tap

Fig. 3-13. System calculations for determining signal levels from head end to last tap.

Designing the Head End

The calculation of expected signal strength at any given receiving site is a long and involved process. Unless you have a situation where clear line of sight exists between transmitter and receiver, these calculations are often misleading. The difficulty lies in the inability to fully account for the effects of buildings, trees, hills, ground conductivity, and weather conditions on the transmission of signals. Moreover many of these signal-affecting things vary from season to season.

DETERMINING SIGNAL RECEPTION CONDITIONS

For practical everyday MATV work, a more reliable means of determining signal conditions is needed. Experience is by far the best teacher. Records of signal strength and picture quality from previous jobs are invaluable aids to estimating antenna requirements for new work in nearby areas.

If you are new to an area, experienced or not, there is nothing that will substitute for an on-site signal survey. The minimum survey equipment needed includes a moderate gain broadband antenna, at least 20' of mast, a calibrated field strength meter (FSM) and a properly aligned portable color TV receiver. Some contractors go as far as equipping a van with full surveying equipment including a 40' crank-up tower with rotator and auxiliary power. Whatever method you choose, you cannot adequately design an antenna system without first knowing the signal conditions in the job site area.

A small antenna of the collapsible type with six to eight elements is usually adequate for field testing. Select one that is lightweight, easy to handle and inexpensively replaced if seriously damaged. Uusally the next to smallest in any supplier's line will do a good job. Get the manufacturer to provide you with average gain characteristics if they are not published. Typically they will run 1 to 2 dB gain for low band and 4 to 6 dB for high band. If the antenna is an all-channel design, the UHF will have 5 to 9 dB gain.

When testing an area for signal, record the readings per reference level (in dBmV) for each channel. For uniformity, it would be a good idea to subtract the gain of the antenna from the readings for all channels involved. This will produce a uniform record of signal strength as would be expected from a dipole on each channel. When it comes time to select an antenna for the system design, you can add the gain of the new antenna directly to those recorded levels to get a good idea of operating levels.

When making a field survey, tune the FSM slowly through its entire range. Make a record of any possible interference signals that may be noted near the desired TV channels. Of special interest are strong FM stations or any other signal that is equal to or stronger in level as compared with the TV signal. Such extraneous signals are potential sources of interference and may require traps or filters to eliminate.

Look at the pictures with a TV set. Don't be too concerned about somewhat snowy pictures because you will cope with low signal level in the antenna system design. Of primary concern is interferences such as ghosts and manmade noises, like power line noise, that won't show on a meter. These too should be recorded and will influence your design. If ghosting is noted, try moving the antenna to see if a ghost-free spot exists nearby. If electrical noise is noted, watch for a while to see if it's steady or intermittent. It could be electrically powered construction tools that will go away when the building is completed. The tracking down of trouble and recommended solutions are discussed in Chapter 7.

SELECTING THE PROPER ANTENNA

The selection of antennas for an MATV system is dependent upon a number of things. Reception of snow- and interference-free signals on all channels is the prime objective. Also important to this selection is the consideration of cost and durability. Minimum signal requirements were discussed in Chapter 1. Coping with interference is covered in Chapter 7.

On small jobs, cost is quite important. For these cases it is common to use broadband antennas. Regardless of job size, if signal conditions are good, the broadband antenna will do a creditable job. Where signal condtions vary from channel to channel, the wisest choice is to use individual antennas for each channel. Individual antennas give you maximum flexibility to cope with interference and to balance signal levels before combination into the system.

It is recommended that Yagis be used for each channel. Most manufacturers offer a line of ruggedized Yagis cut for each of the VHF channels. Various heavy-duty models are offered for UHF channels and selection should be made according to gain requirements. Heavy-duty antennas offer the long life and durability under all weather conditions that are recommended in all MATV work.

RECEIVED SIGNAL AND ANTENNA HEIGHTS

The signal strength received at any given site is dependent upon frequency, transmitter power, distance and intervening terrain. All of these signal-affecting things are not under our control. The only variable usually open to us is the height above ground at which the antennas can be mounted.

As a rule of thumb, if the antenna height above ground is doubled, you can expect up to 6 dB of signal increase at VHF. At UHF even greater gains can be realized, up to 12 dB. This rule of thumb holds fairly true for long distance reception where the receiving site is not in a direct line of sight with the transmitter. Antenna height variations in line of sight conditions will have very little effect on signal level.

Where height variations do affect signal strength, the rule of thumb is based on height above average terrain. If you locate the antenna on top of a building, doubling the antenna height means doubling the building height.

Stacking Antennas for Gain

When stacking antennas for gain it is necessary to follow a few basic rules. Fig. 4-1 illustrates these requirements. The vertical spacing between antennas should be no less than 0.67 wavelength. This is a minimum distance. Wider spacing than this is permissible but no change in gain performance will

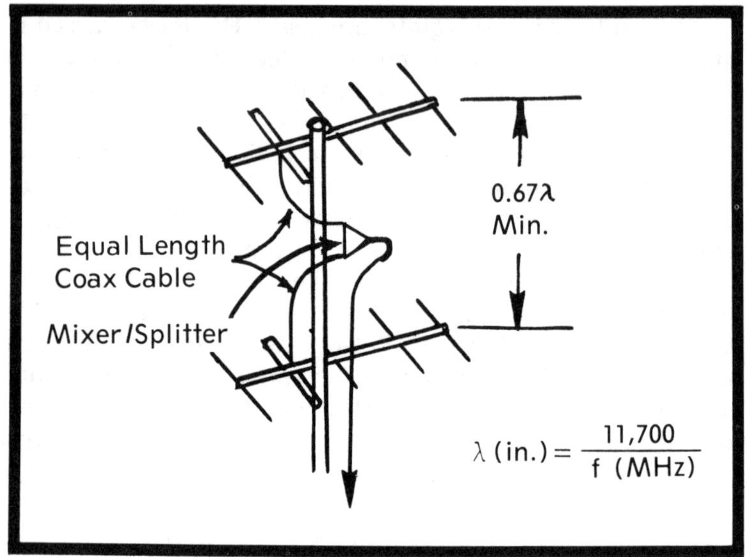

Fig. 4-1. Recommended method of stacking antennas for gain. Minimum vertical spacing in inches should be calculated from the formula given.

result. The wavelength for each channel is given in Table 3 of the appendix or may be calculated by the formula shown in Fig. 4-1. If the antennas to be stacked are single-channel Yagis, the wavelength distance is calculated at the picture carrier frequency. If the antennas are broadbands, the distance is determined by the lowest frequency channel to be received.

Many forms of interconnecting stacked antennas have come and gone. One best method remains. This method is the use of a splitter in reverse. A hybrid splitter is essentially a directional coupler that divides the signal power into two equal outputs. Used in reverse, under the proper conditions, it can combine two equal signals to be 3 dB or double the power of each input. Since two antennas receive essentially equal strengths, maximum increase will be realized. Proper conditions require two things:

1. In-phase reception of signals by both antennas. This can be achieved by mounting them on a common mast.

2. Assuring in-phase addition of signals within the mixer by using equal length cables from each antenna. The exact length of each cable is not important as long as they are both equal to within ½ inch.

Since this method of combining is not frequency-sensitive, it can be used to combine broadband antennas to realize a 3 dB improvement on all received channels.

If still more signal is needed, four antennas may be stacked for additional gain. The total gain of a four-antenna array would approach 6 dB above that of a single antenna. Fig. 4-2 illustrates the four-bay stack. Two acceptable methods of combining are shown.

The laws governing stacked antennas state that you will get a maximum of 3 dB increased gain by doubling the number of bays in an array. In reality you will only get slightly more than 2.5 dB increase because the mixer-splitter is not loss-free. Four antennas in array represents a practical limit for MATV work. To get another 3 dB would require going to eight bays. This becomes a mechanical monster and is at the point where you must seriously question the value of the increased gain vs the expense and complication.

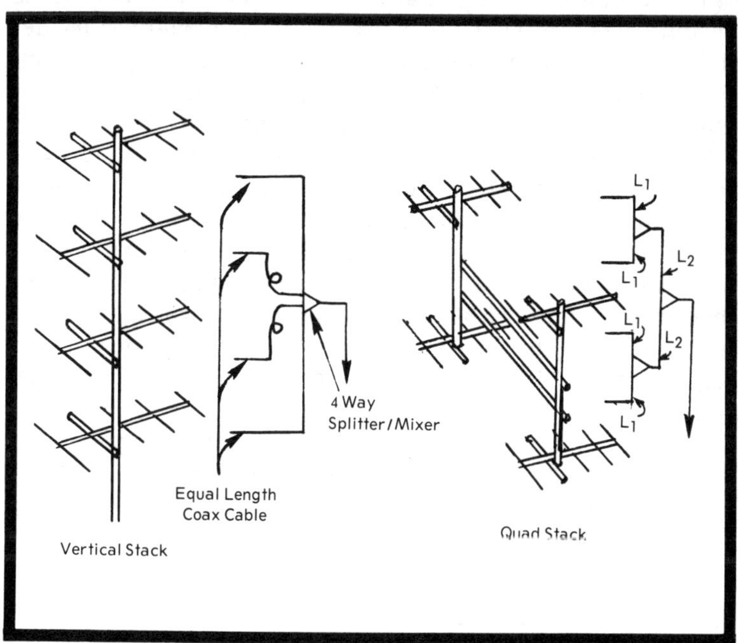

Fig. 4-2. Four antennas may be stacked vertically or as a quad stack. Note that all lead-in cables must be of equal length to assure in-phase mixing of signals. Minimum vertical separation distance is 0.67 times wavelength. Minimum horizontal separation should provide mechanical clearance between antennas.

Stacking Antennas Against Interference

Reception of distant TV signals is often hampered by co-channel interference. This interference produces closely spaced, horizontal dark lines on the screen, something like viewing the picture through venetian blinds. It is caused by the reception of weak signals from some other station operating on the same channel as the one you're trying to receive. This interference can be received from co-channel stations at any distance but usually it is limited to within 200 miles. It is typically an intermittent type of interference but can become quite severe during adverse weather conditions. The most effective way to cope with co-channel is to use a horizontally stacked antenna array.

Fig. 4-3 illustrates the array configuration for a horizontal co-channel stack. The two antennas are aimed at the desired station as indicated by the incoming wavefront. The undesired signal is shown arriving at angle A. The horizontal spacing of the antennas is adjusted to cause a half wavelength or 180 deg difference between antennas to the undesired signal. This arrangement gives a 3 dB increased gain to the desired signal while the undesired signals are self-canceling.

To design such an array, it is necessary to determine angle A quite accurately. For this purpose it is recommended that that you use a sectional aeronautical map of the area. Television station towers are shown on these maps. Locate your position accurately and measure the angle between the desired and undesired stations. If the angle is less than 90 deg, apply it directly to the formula. If the angle is greater than 90 deg, subtract it from 180 deg and apply the result to the formula.

By way of example let us figure the horizontal spacing for a co-channel array for an angle A of 160 deg for Ch. 7:

180 - 160 = 20 deg

1. $H\lambda = \dfrac{1}{2 \sin 20 \text{ deg}}$
2. $H\lambda = \dfrac{1}{2 \times 0.342}$
3. $H\lambda = \dfrac{1}{0.684}$
4. $H\lambda = 1.46$

Ch. 7 Pix = 175.15 MHz

5. $\lambda \text{ in.} = \dfrac{11{,}700}{175.15}$ (in air)
6. λ in. = 66.8 in.
7. H = 66.8 x 1.46
8. H = 97.4 in.

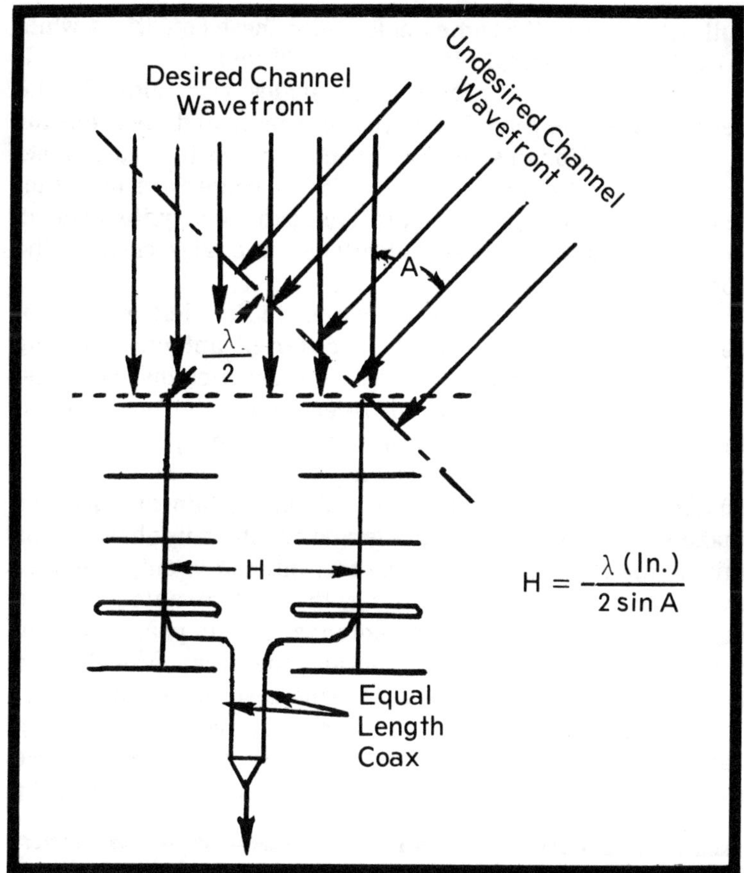

Fig. 4-3. Horizontal stacking of two or four antennas may be adjusted to reject interference from any specified angle. Horizontal spacing in inches is determined by the formula given. NOTE: If the formula requires a spacing that will cause the elements of the antennas to overlap, use a spacing of 3H to achieve the same results.

It doesn't matter which side the co-channel signal comes from because this array will produce **four** nulls in its polar pattern. A null will occur at 20 deg left and right of straight forward and 20 deg left and right of straight back. At wide angles between 45 and 90 deg, you will find that the formula calls for spacings quite close together. If you mount the antennas on a tower you may find that the spacing is so close that the antenna elements touch or are uncomfortably close to the support structure. In these cases you may use any **odd** multiple of the calculated distance for H such as 3 H or 5 H. A null will occur at the same angle under these conditions while providing mechanical clearance for the antennas.

If you apply this formula to angles very close to the desired direction you will find that impractical dimensions for H will occur. For this reason it is considered that co-channel angles of less than 5 deg either side of center are not practical. Even at 10 deg. the separation distance is quite large on low band channels. This is also true for angles close to the back of the antennas.

For interference reception within 10 deg. left or right of being directly from the back, a different antenna stacking technique can be employed. Fig. 4-4 illustrates the technique of **vertical stagger stacking** for rejection of interference from the rear. Consider first the desired signal coming into the front of the antenna array. Signals will arrive at the top antenna a quarter wavelength earlier than at the bottom antenna. To make certain that these signals add together in phase at the mixer, the coax cable from the top antenna is cut a quarter wavelength **longer** than that for the bottom antenna. This delays the signal from the top antenna to be in phase with the signal from the bottom antenna and the array provides the typical 2.5 to 3 dB additional gain to the desired channel.

Signals from the rear will arrive at the top antenna a quarter wavelength later than at the bottom antenna. These signals will also be delayed another quarter wavelength by the

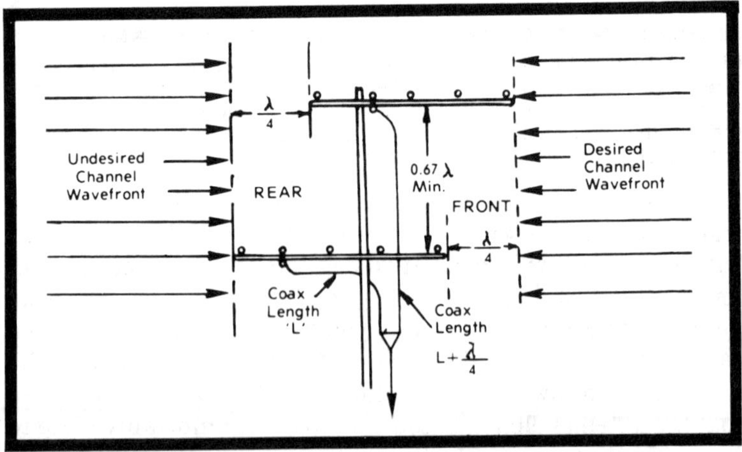

Fig. 4-4. Vertically stacked antenna array, staggered to improve the rejection of unwanted signal reception from the back.

extra length of cable running to the mixer. At the mixer these signals will be a half wavelength or 180 deg out of phase and will be self-canceling.

When designing such an array be sure to use proper dimensions for wavelength. The formula for determining wavelength in air was given in step 5 of the previous example. This is the proper formula for calculating the stagger distance for mounting the antennas. For determining the proper dimension for the extra cable length, you must apply a factor for the type of cable used, called **the velocity of propagation factor**. This is because rf travels slower in cable than in air and a wavelength in cable is physically shorter than in air. This factor is usually specified by the cable manufacturer and is typically 0.66 for solid dielectric cables and 0.82 for foam dielectric cables.

The following steps show the procedures for determining the dimensions of a stagger-stack at Ch. 2:

1. $\lambda \text{ in.} = \frac{11,700}{55.25}$
2. λ = 211 in. in air
3. 211 ÷ 4 = 53 in. staggering dimension
4. 211 x 0.67 = 141 in. vertical antenna spacing.
5. 211 x 0.82 = 172 in. one wavelength in foam dielectric cable
6. 173 ÷ 4 = 43 in. of extra cable for the top antenna lead-in

The horizontal stack and the stagger-stack antenna array can be counted on to provide minus 20 dB of rejection to unwanted signals. Carefully built arrays can easily exceed minus 30 dB rejection. These arrays can also be used to fight interferences other than co-channel. If you can determine that a ghost is being caused by a reflection from a known location, these arrays will be quite useful.

FILTERS AND TRAPS

The need for filters and traps in the head end design is primarily determined by the site signal survey. The choice of using a trap or filter depends on what you want to get rid of. Broadly speaking, a trap is used to eliminate a single interfering frequency, a filter is used to reject a group or band of possible interfering frequencies.

The most common use for filters is on single-channel antenna installations. Each antenna receives its assigned channel plus something of the other channels present. These are considered unwanted signals that must be removed before the signals are combined. If for instance a Ch. 7 Yagi received some Ch. 9 signal that was allowed to mix with the signal from the Ch. 9 Yagi, you can expect poor picture quality. This is because the Ch. 7 Yagi is a poor antenna at Ch. 9's frequency and it will distort the signal badly. This distorted signal mixed with a good signal will produce smearing, ghosts, poor color, or loss of color. Simply filtering each antenna to reject all unwanted signals will prevent this situation. The filters could be separate units, antenna combining networks, or single-channel amplifying devices such as preamps.

Filters that reject a specific band and pass all other frequencies are also available. Common in this type is the FM band reject filter to prevent overloading of amplifiers with strong FM signals. Crossover filters or band separation filters can also be used to reject unwanted signals. Many manufacturers make such a unit for closed circuit systems that use the frequencies below Ch. 2 for closed circuit work. This filter splits the 54 to 890 MHz TV bands from the so-called **subchannel band**, dc to 47 MHz. Such a filter could be used to pass only the TV band from the antenna to the system while any low frequency interference, such as CB radio, etc., would be filtered out.

Traps are used to reject single frequencies because they are narrow-bandwidth devices that may be tuned to specific frequencies. This unique ability makes them particularly suited to solving difficult interference problems. It is often desirable to receive a rather weak distant channel because of the programming offered. Too often, there is a strong local adjacent channel that makes this difficult if not impossible. Typically a bandpass filter will do no good in this situation because the skirts of such filters are too broad. A high Q tunable trap will provide a reliable minus 40 dB attenuation to the unwanted signal.

Tunable traps may be needed to reject other single-frequency interferences such as FM, conversion beats, or other spurious signals. Traps are field tunable with the aid of an FSM as discussed in Chapter 7.

UHF AND THE CONVERTER

Reception at UHF is not only subject to all the same rules that apply to VHF but also to a host of other considerations.

Antenna gains are usually higher, cable loss is higher, and greater amplifier gains are usually required. We saw in the system layout problems of Chapter 3 that UHF channels could be carried directly on small- and medium-sized systems. In large systems, it is wise to consider conversion to VHF.

Conversion of UHF makes the design and operation of large systems much easier than on-channel distribution. It does, however, introduce a number of potential problems which the designer must consider. Conversions must be beat-free and must not interact with each other in multi-UHF channel areas.

Equipment manufacturers will advise you about prohibited conversions with their equipment. Certain U to V conversions are not recommended because of the probability that a beat will result that cannot be trapped out. Consider Fig. 4-5 which shows the essential elements of a converter. The desired UHF channel enters through a band pass filter where it is mixed with a locally generated signal in a nonlinear device like a diode. These two signals beat together to produce

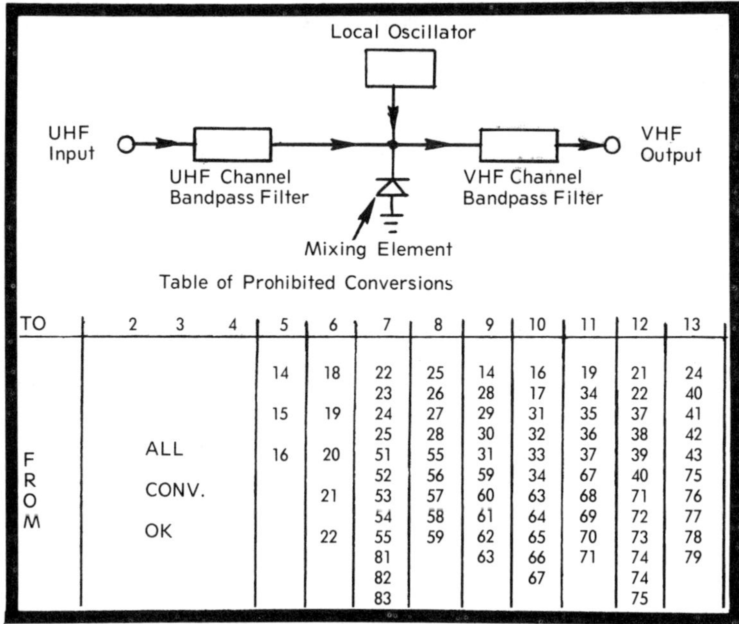

TO	2 3 4	5	6	7	8	9	10	11	12	13
		14	18	22	25	14	16	19	21	24
				23	26	28	17	34	22	40
		15	19	24	27	29	31	35	37	41
				25	28	30	32	36	38	42
F	ALL	16	20	51	55	31	33	37	39	43
R				52	56	59	34	67	40	75
O	CONV.		21	53	57	60	63	68	71	76
M				54	58	61	64	69	72	77
	OK		22	55	59	62	65	70	73	78
				81		63	66	71	74	79
				82			67		74	
				83					75	

Fig. 4-5. Block diagram of UHF to VHF converter showing essential elements. The table lists all so-called prohibited conversions. These conversions will produce beat interferences that fall within the output VHF channel that cannot be trapped or filtered out.

many other frequencies such as harmonics of each frequency, sums and differences of the frequencies, etc. The difference frequency, UHF channel minus the local oscillator (LO), is made to fall at the desired VHF output channel and therefore passes to the output terminal. The other generated beats are not permitted to get out of the converter due to the action of the filters. See Fig. 4-6, which illustrates typical converters.

Consider the choice of Ch. 41 to Ch. 13 as a conversion. Channel 41's picture is at 633.25 MHz. The LO frequency required would be 422 MHz. And 633.25 minus 422 equals 211.25 or Ch. 13's picture carrier. The problems with this are many. Since Ch. 13 is produced within the converter, so is the second harmonic of Ch. 13 which happens to be 422.50 MHz. Now we have a second frequency present that will mix with the incoming Ch. 41 and produce a 500 kHz beat in the Ch. 13 output. Worst of all is the fact that the second harmonic of the LO is generated, at 844 MHz. This will mix with the Ch. 41 signal also. And 844 minus 633.25 equals 210.75 which is an unwanted signal 500 kHz below Ch. 13 picture carrier. This illustrates just a couple of reasons why various U to V combinations are forbidden. See Fig. 4-5, which lists all such prohibited conversions.

Even when permissible conversions are used, it is easy to get into trouble with beats. This can occur when more than one channel of UHF is to be converted. Assume the reception of Ch. 27 and 29. A natural thought would be to convert these to Ch. 7 and 9 if unused in that area. **This could be real trouble**! A Ch. 27 to Ch. 7 converter will also produce an incidental conversion of Ch. 29 to Ch. 9. This will happen because the filter response curves in converters are not sufficiently sharp to reject signals only two channels away. An incidental conversion of 27 to 7 will also occur through the 29 to 9 converter. When these converted signals are mixed together, **each channel will contain an unwanted beat**. If you must make a conversion such as this, include a high quality VHF channel pass filter at the output of each converter to help reject the unwanted incidental conversions.

If you face the above reception problem, a better choice of conversions would be 27 to 13 and 29 to 7. Incidental conversions will still occur but they will fall well outside the band of interest where they can do no harm.

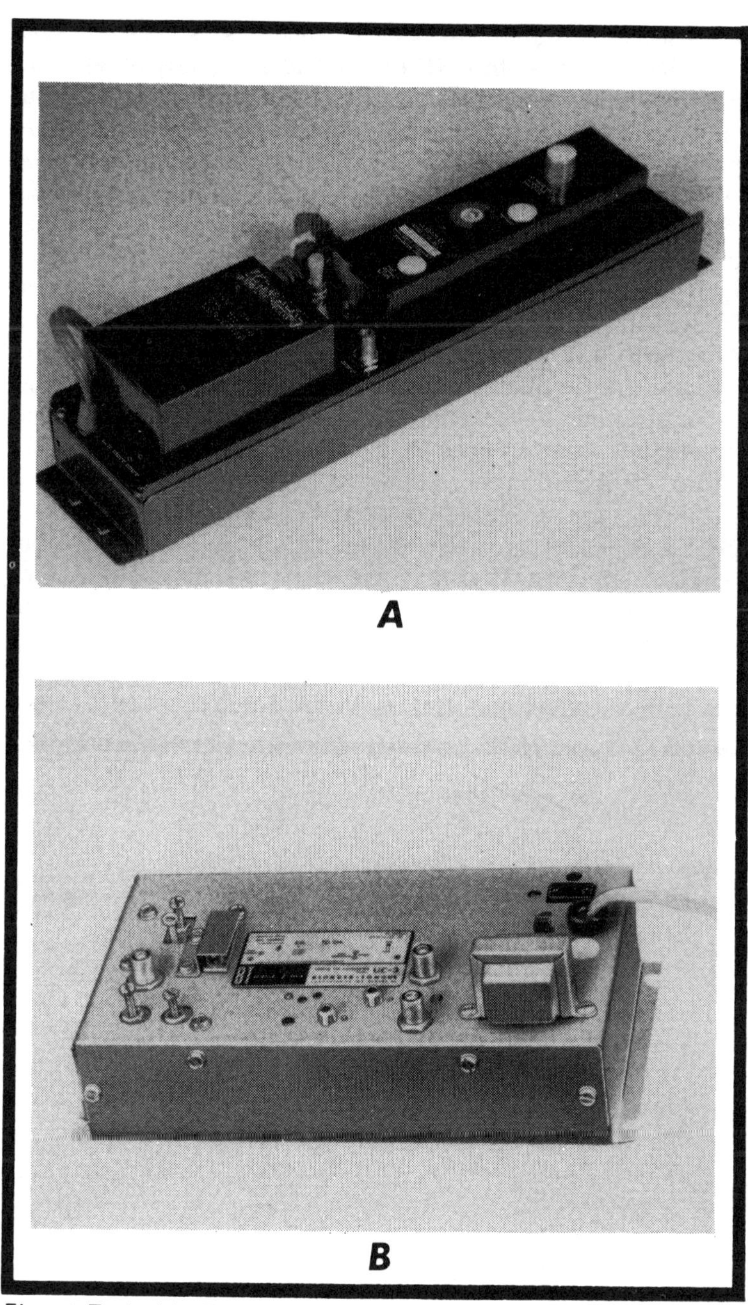

Fig. 4-6. Typical UHF to VHF converters. A is Model DA-UVC, courtesy of Winegard. B is Model UC3, courtesy of Blonder-Tongue.

Another problem to worry about is **self-conversion.** Consider an area with UHF Ch. 17 and 29, or any other combination of 12-channel spacing, which is common. Here's what could happen if you should choose to convert 29 to 4. The Ch. 29 picture carrier is at 561.25 MHz. The LO frequency would be 494 MHz. Now, 561.25 minus 494 equals 67.25 MHz for proper conversion to Ch. 4. If the Ch. 17 signal received by the Ch. 29 antenna is strong, enough could get through the converter's input filter to mix with the Ch. 29 signal. The Ch. 17 sound carrier is at 493.75 MHz. Since 561.25 minus 493.75 equals 67.5 MHz, there's a 250 kHz beat inside Ch. 4. Obviously this problem can be avoided by selecting a different conversion.

When planning to convert UHF to VHF, take time to check for possible interferences that could be caused by an unfortunate combination of otherwise permissible conversions. Check Fig. 4-7, a simple frequency scale with the VHF and UHF stations of an area shown. Scale A is one possible selection of U to V conversion that has the potential of producing beat interferences. Scale B is a different selection of conversions that is free of potential trouble. After locating the VHF and UHF channels, make an arbitrary selection of one U to V conversion. Calculate the LO frequency by sub-

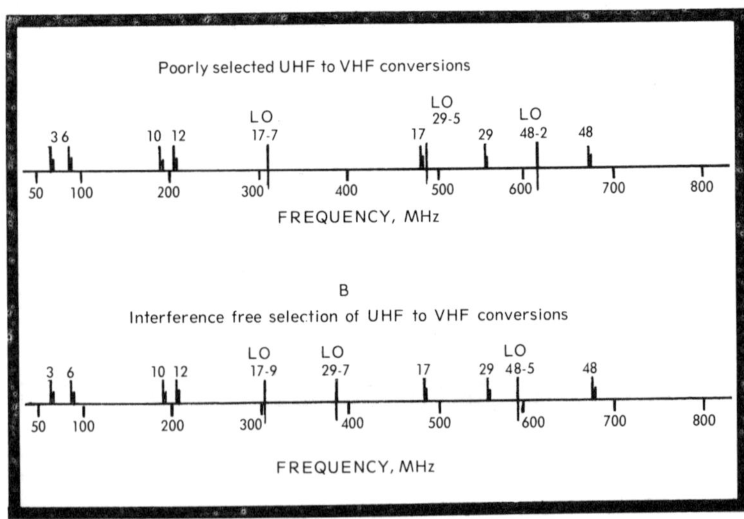

Fig. 4-7. Frequency spectrum, 50 to 800 MHz, showing location of channels and local oscillator frequencies for (A) poorly selected UHF to VHF conversions and (B) interference-free selection of conversions. See text for explanation.

tracting the converted-to-VHF picture carrier frequency. In scale A, Ch. 29 to 5, the LO is 561.25 minus 77.25 equals 484.00 MHz. Plotting this on the scale shows immediately that the LO falls within the band for Ch. 17. This has the potential of producing trouble and a different conversion should be selected for Ch. 29 such as 29 to 7, as in scale B.

Again, looking at Fig. 4-7, you can see that Ch. 28 to 2 is a potential problem conversion, when the LO is plotted as in scale A. Note that the spacing between the LO and Ch. 29 and 48 are about equal. This is a dead giveaway to potential image frequency problems and should be avoided.

This is not a prediction that such problems will actually occur. Should the conversions as selected in A be installed, trouble may or may not exist. It all depends on filter design within the converters, operating signal levels, and antenna connections. By going through this type of exercise, however, you can easily **avoid** the possibility of trouble.

MASTER AMPLIFIERS AND PREAMPS

In Chapter 3, you calculated the system requirements for head end output level required to deliver the desired signal to all outlets. Selection of the right master amplifier is primarily dependent upon those calculations. Generally speaking, if a system requires less than 60 dBmV out of the head end, a broadband amplifier will usually suffice. An exception to this would be when you have widely fluctuating antenna signals that call for agc strips. For signal levels over 60 dBmV, strip amps are the best bet.

It is generally wise to select an amplifier that has about 3 dB more output capability than calculations require. This will give you a margin of safety that is reasonable but not prohibitively expensive. Safety margins are desirable in every system but should only be calculated into the design at one place, **here**. This is the place where it will do the good expected of a safety margin and you will know just how much it will cost by comparing amplifier prices.

The following examples will explore the techniques of selecting head end equipment. The 80-outlet system calculated in Chapter 3 and illustrated in Fig. 3-1, calls for a minimum signal level out of the head end of 35.9 dBmV. This should be raised to approximately 39 dBmV to include a safety factor.

When consulting various manufacturers' catalogs you will find many amplifiers for small systems, with output ratings between 40 and 45 dBmV. These amplifiers are inexpensive and any one of them will do a creditable job. Gain specs for these amps range between 22 and 30 dB. This tells you that the input signal from the antenna must be from about 9 to 17 dBmV. In metropolitan areas this level of signal is easily achieved with a modest antenna. Signal levels are usually quite uniform from channel to channel and a single amplifier will do the job.

If antenna signal levels are weak, say 0 to minus 6 dBmV, you might consider a preamp. Don't overlook the possibility of serving the entire system from such a preamp. Most MATV preamps have high gain, approximately 30 dB, and good output capability in the 40 dBmV region. Assume the following channels and dipole reference signal levels.

	Ch. 3	Ch. 6	Ch. 10	Ch. 12
Dipole ref. level	16 dB	15 dB	12 dB	11 dB
Ant. gain	6 dB	6 dB	10 dB	10 dB
	———	———	———	———
Expected level	22 dB	21 dB	22 dB	21 dB
Required output	39 dB	39 dB	39 dB	39 dB
	———	———	———	———
Gain required	17 dB	18 dB	17 dB	18 dB

A broadband antenna was chosen with higher gain on the high channels than on the low ones to provide equalizing of signals above the test levels. Subtracting the expected signal levels from the calculated system requirement shows the need for 17 to 18 dB gain in the head end amplifier. If you select even the lowest gain units available, 22 dB, the system could overload from too strong an input. Adding a minus 6 dB **loss pad** to the antenna lead-in will balance things nicely.

The system requirements of Fig. 3-4 calculated out to be 38 dBmV. Adding the 3 dB safety factor means 41 dBmV. Again the selection of the master amplifier is based on output requirements. Assume the following channels and dipole reference levels.

	Ch. 3	Ch. 6	Ch. 10	Ch. 12
Dipole ref. level	6 dBmV	1 dBmV	–5 dBmV	–11 dBmV
Ant. gain	8 dB	8 dB	8.5 dB	8.5 dB
Expected level	14 dBmV	9 dBmV	3.5 dBmV	–3.5 dBmV
Required output	41 dBmV	41 dBmV	41 dBmV	41 dBmV
Gain required	27 dB	32 dB	37.5 dB	44.5 dB

There is a rather wide variation between channel levels and each channel must be handled individually. Individual antennas give this flexibility. The master amplifier can again be broadband and must have an output capability of at least 41 dBmV. Assume a gain of 30 dB and follow the requirements for each channel on Fig. 4-8. As you see, Ch. 3 mixes straight into the amp through a mixing network. But Ch. 6 is shy of having enough signal by 4 dB. If a higher gain antenna could be found it could save using a preamp. To be assured proper operation a preamp should be used with most of its gain thrown away by padding. Both Ch. 10 and 12 need preamps and appropriate pads to balance the levels before they enter the master amplifier.

The all-channel system as calculated in Chapter 3 calls for a head end output of 47.5 dBmV minimum. With safety factor this can be rounded to 50 dBmV. This system also calls for UHF distribution which is handled in the same way as for VHF only. Separate UHF and VHF amplifiers may be used or one of the high gin combination U-V amps available will meet this requirement. Fig. 4-9 shows a typical head end setup using individual VHF and UHF master amplifiers. Typical signal

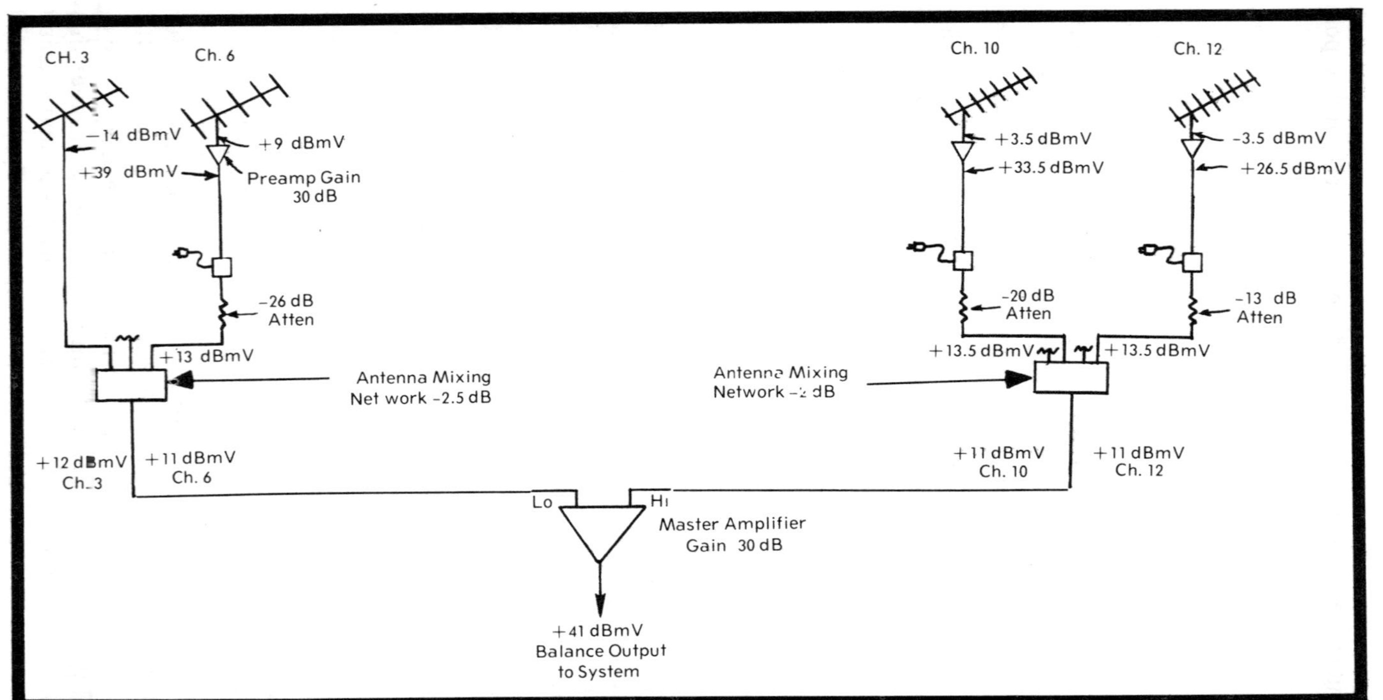

Fig. 4-8. Unequal signal levels from different channels are balanced by a combination of preamps and attenuator pads.

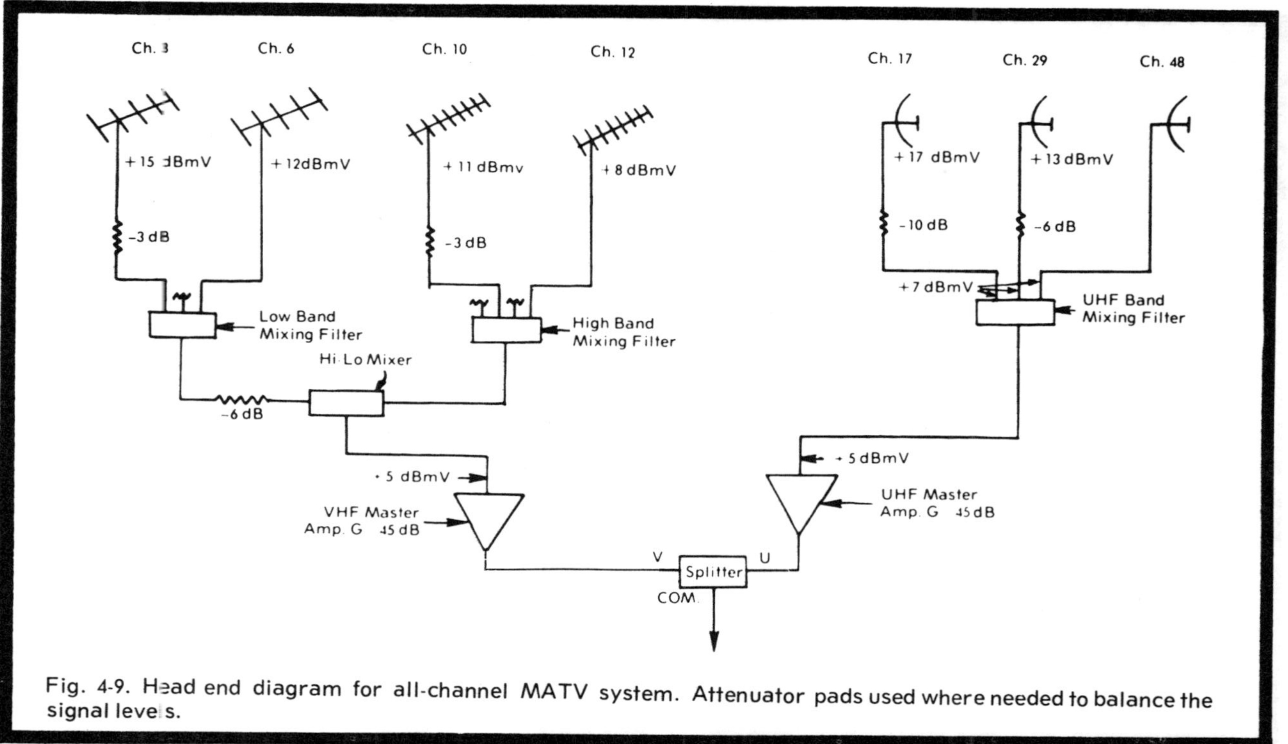

Fig. 4-9. Head end diagram for all-channel MATV system. Attenuator pads used where needed to balance the signal levels.

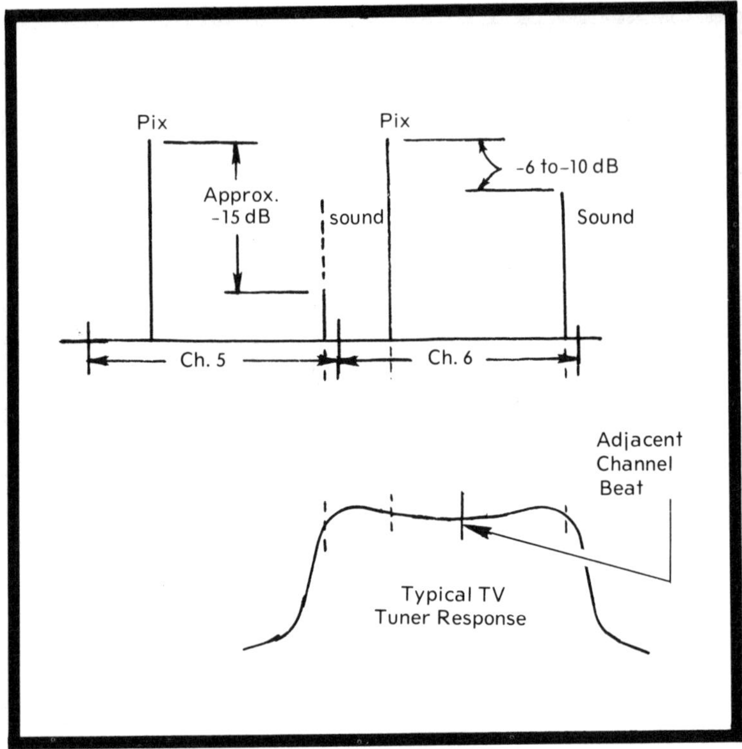

Fig. 4-10. Beats are produced in adjacent channel systems because lower adjacent sound carriers can get through the tuner of TV sets. Attenuation of sound carrier levels eliminates this type of interference.

levels are shown at key points throughout the layout for ease of understanding. If you make it a habit to mark all operating levels on your system designs, they will become an invaluable aid to quick installation, setup, and troubleshooting.

ADJACENT CHANNEL OPERATION

The need to distribute adjacent channels comes up quite often in MATV system design. This is typical in systems such as the trailer park layout of Fig. 3-9. The system is sufficiently large to require line reamps even when figured for VHF-only distribution. With UHF in the reception area, conversion to VHF often gives you an adjacent channel situation. Such a situation is recommended in the UHF conversion scheme of Fig. 4-6B. Here, adjacent channel operation occurs on Ch. 5 and 6 and on 9 and 10.

The reason that adjacent channels produce any problems is a function of the TV sets themselves. The standard allocations of TV channels allows for a guard band of at least one channel between stations broadcasting in a given metropolitan area. This makes it easy for the TV receiver to cope with a wide variation of signal levels without interference from other strong nearby signals. Most receivers contain adjacent channel traps in the i-f section to trap out any stray signals from distant stations.

In cable systems, all channels are adjusted to equal levels and many TV receivers cannot satisfactorily reject the relatively strong signals of an adjacent channel. The problem occurs in the tuner before the signal gets to the adjacent channel traps in the i-f strip. See Fig. 4-10, which shows the spectrum of adjacent Ch. 5 and 6 with the typical response of a TV tuner, tuned to Ch. 6. You can see that a considerable amount of the Ch. 5 sound carrier will beat with the picture carrier in the tuner mixer to produce a beat at 1.5 MHz inside Ch. 6.

The FCC requires all broadcasters to operate their stations with the sound carrier between minus 6 and minus 10 dB below their picture carrier level. This is shown on the Ch. 6 sound carrier. If Ch. 5 were placed on the cable as received from the station, shown in dotted lines, the beat produced would be strong enough to see on the TV screen. It has been found that if you reduce the level of the sound carrier to minus 15 dB below the picture carrier, the beat is reduced below the level of visibility. This does not affect the quality of the sound and most sets could stand another 5 dB of sound level reduction before sync buzz or noise were noticed.

This sound carrier level reduction can be done with special traps available from MATV equipment manufacturers. See Fig. 4-11. Sound carrier reduction must be done in all adjacent channel systems but it is only necessary on the lower adjacent channels. Don't worry about variations in sound carrier level from channel to channel. The sound carrier is FM and audio volume is not affected by variations in carrier level.

In Fig. 4-12 is a typical example of a single-channel, high level output head end. As in all systems, the type of equipment is selected per output requirements. Preamps are used to provide additional gain for weak signals as in previous examples. This example does combine many of the requirements of good head end engineering that deserve to be pointed out.

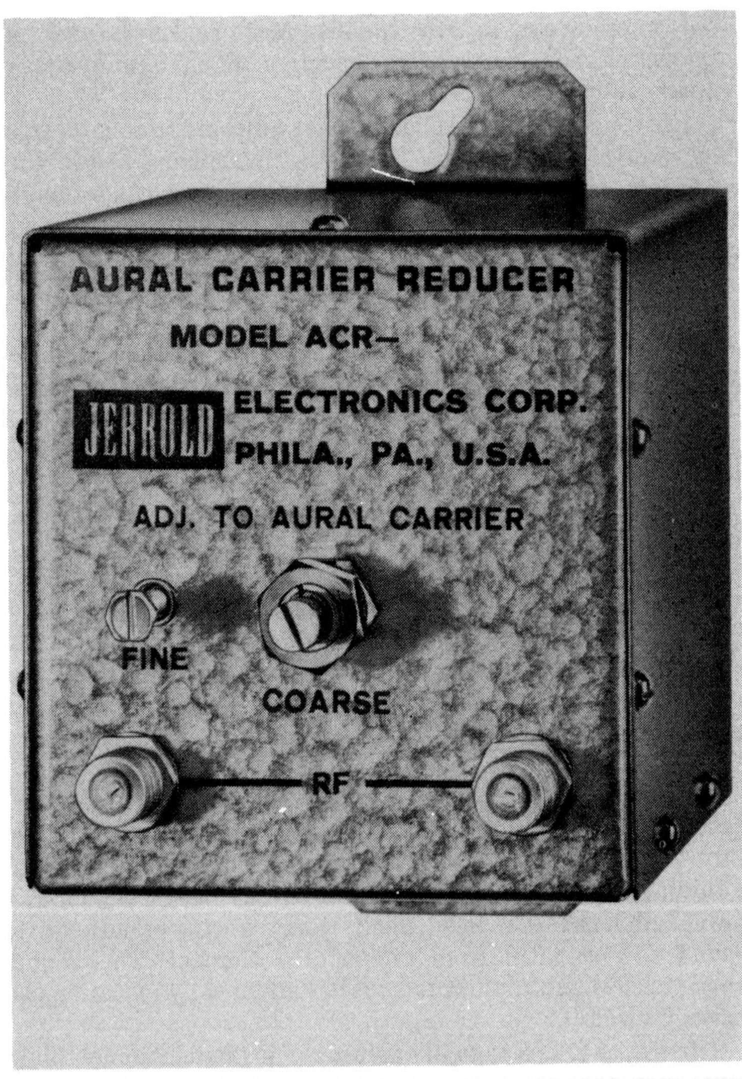

Fig. 4-11. Special trap used to reduce the level of TV channel sound carrier without degrading picture or color quality. Model ACR-25, courtesy Jerrold Electronics Corp., tunes Channels 2 through 5. Other models tune Channels 7 through 12.

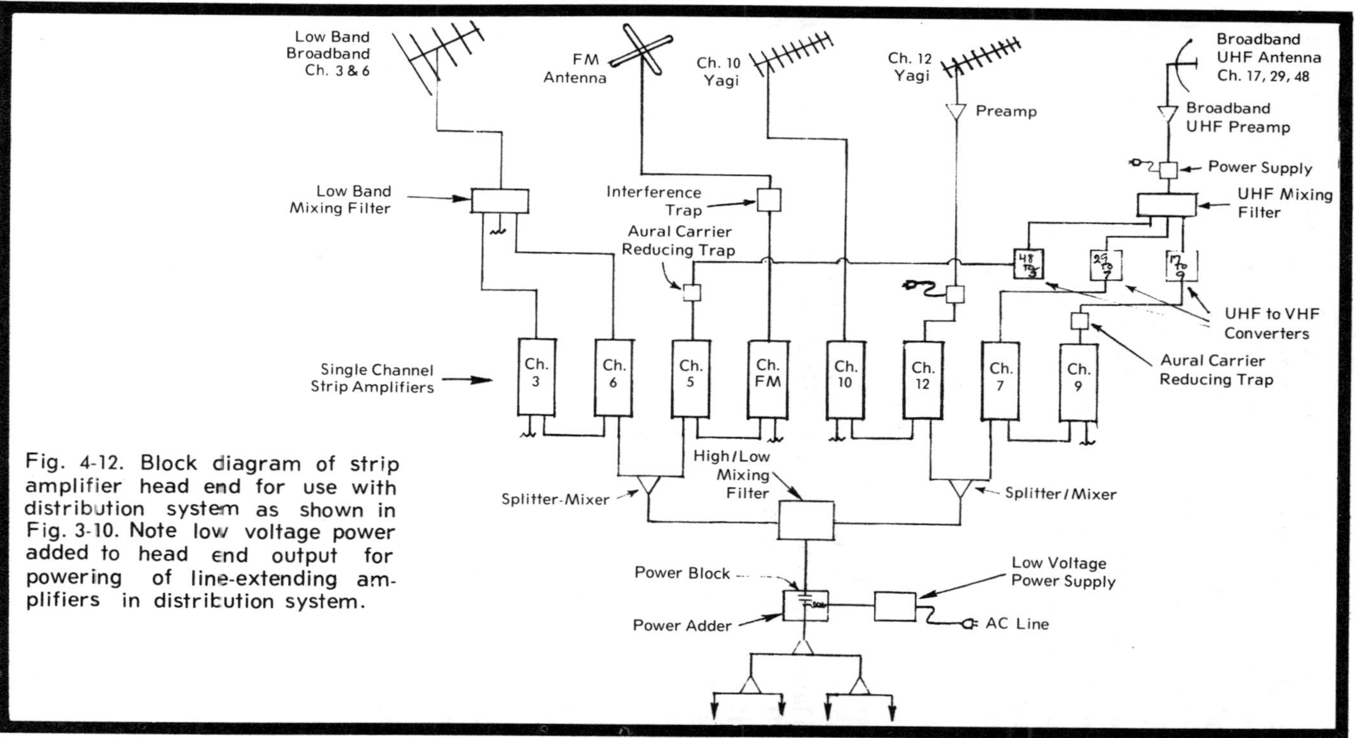

Fig. 4-12. Block diagram of strip amplifier head end for use with distribution system as shown in Fig. 3-10. Note low voltage power added to head end output for powering of line-extending amplifiers in distribution system.

Local Ch. 3 and 6 are received from a common low band, broadband antenna. Channels are separated by using a low band mixing filter in reverse. Each channel is then fed to individual channel strip amps. An omnidirectional receiving antenna is used for full area coverage of the FM band. Note that an interference trap is included in the FM antenna down-lead. The purpose of this trap is to eliminate any unwanted Ch. 6 reception through the FM antenna that could get through the skirts of the FM strip. If this were allowed to happen, the unwanted Ch. 6 energy could distort the picture quality when mixed together in the output.

Channels 10 and 12 are straightforward in design with preamplification included as desired. The UHF Ch. 17, 29, and 48 are received on a common antenna. A preamp is used for two reasons. If the signals are below 10 dBmV and you have a long down-lead, the preamp will overcome loss of quality due to large value cable losses at UHF. Secondly if you convert to VHF as in this example, it is desirable to hit these converters with a high signal level, 20 dBmV min., because most converters have high noise figures. The conversions selected are those previously discussed.

In this head end arrangement note that converted Ch. 5 and 9 fall as lower adjacent to VHF Ch. 6 and 10. To assure beat-free operation of the system under these adjacent channel conditions, the system must include sound carrier reduction traps for Ch. 5 and 9. Since no other adjacent channels exist, there is no need for any more sound level reducing traps.

Individual channel strip amps are used for each channel plus FM. Each channel can be controlled individually and all should be set at output levels slightly higher than minimum system design requirements and well within the rated maximum output. Each strip amp is equipped with a means of mixing signals, usually by the loop-through method. This output circuit involves filters within each unit which pass the amplified signal to one or both output terminals. These filters must also cope with the loop-through signals of other amplifiers without serious distortion or attenuation. If all strips were loop-connected in Fig. 4-12, serious distortion would occur. Signals from the most distant amp to the combined output would suffer a small loss, under 0.5 dB, as it passes through each combining circuit. With seven amplifiers as in this example, the total loss could approach 3 dB. Worse than that, however, is the distortion of signal quality on Ch. 5, 6, 9, and 10 as each pass by the highly reactive filter skirts of each other's adjacent channel.

Both of these undesirable things can be eliminated by interconnecting the outputs as shown in the illustration. Note that only nonadjacent channels within the low and high bands are looped together. Adjacent groups within each band are combined by using a standard splitter in reverse. This isolates the adjacent channel outputs by virtue of the directional coupler design of hybrid splitters. **Be careful to include the 3.5 dB loss of these splitters in the system calculation of output requirement.**

Combined high band and low band groups are then mixed with a high-low mixing filter to provide all channels onto a common line. To cope with power requirements of the system layout, Fig. 3-9, a low voltage power supply is connected into this line. Signals are split four ways to meet the system requirements with power being sent to reamps throughout the system.

5 System Installation and Turn-On

Installing MATV systems is much like the installation of telephone or intercom systems. The basic tools needed to do an efficient job are also very similar.

TOOLS AND USAGE

Each technician should have a tool pouch with the usual pliers, screwdrivers, diagonal wire cutters, etc. The specialized tools for the MATV technician are:

1. **Crimping tool**—to crimp solderless cable connectors.
2. **Scissors**—for trimming cable braid wires.
3. A 7/16" open end wrench—to tighten 'F' connectors.

Heavier tools are needed to do roughing in work. Antenna mast installing may mean setting anchor bolts into cement or brick. Head end equipment locations are often in electrical or mechanical machine rooms and must be protected by cabinets installed against cement or block walls. Cables will often be installed in conduits. This requires a conduit fish tape. For installations that are not in conduit, the coax is usually roughed into the walls while they are still open. To anchor the cable in place a staple gun with ⅜" staples does a good job. When using a staple gun, however, be sure not to pierce or crush the cable. Also be sure to leave plenty of slack at tap or splitter locations to make their installation easier when the walls have been finished. Remember, you have to go back and install them yourself.

Very special tools are needed to do underground work or pole line stringing of cables. To gain knowledge as to how to do this kind of work, it is a good idea to consider contracting it out to someone who is equipped to do it. Supervising such subcontracted work will give you a feel for the investment in equipment necessary to do it on your own. If it's an occasional thing, you may just continue to sub out that portion.

ANTENNA INSTALLATIONS

Antennas installed on the roof of a building should be located so they do not interfere with the operation of other facilities on that roof. Air conditioning is usually located on the roof and requires a lot of attention. Your antennas should not be in the way. More important is to avoid close proximity to objects that can interfere with signal reception. Fig. 5-1 shows how four antennas are located by themselves clear from other rooftop items. Note that there is plenty of clearance above the roof for the lower antennas. To avoid building a single structure that would be difficult to service, two masts support the antennas with adequate clearance between any two and within easy reach of a stepladder for maintenance.

Antenna arrays are quite useful for getting higher signal levels from distant stations. These arrays are easy to design

Fig. 5-1. A typical MATV antenna installation. Note the use of separate antennas for each channel. Also note the generous spacing between antennas.

as was seen in Chapter 4. Physically building one of these arrays is quite another thing. The most difficult to handle is arrays of low band channels.

A quad stack of four Ch. 4 Yagis is shown in Fig. 5-2. This array was built on the ground and is being raised into position on top of the tower. Note the heavy structure that makes up the framework that holds the antennas in position. Each structure like this must be custom-built for each job since the dimensions change with channel and desired pattern.

When you install an antenna system, be sure to use materials and hardware that will last. All masting should be heavy wall galvanized conduit, 1½ in. ID. When installed, the mast should be capped on the top end and left open on the bottom. This will keep the inside essentially dry and provide maximum protection against the elements. All mounting hardware should be galvanized or stainless steel. Nuts and bolts of stainless steel are expensive but not prohibitively so. **Do not use cadmium plated hardware.** It looks good but won't last long when used outdoors. This holds even for things as simple as washers. If they rust out, it probably won't seriously hurt the installation but the rust will wash off during every rainstorm and stain the side of the building. This is ugly to see and certainly won't qualify as an installation you would want to show off.

When considering an antenna installation on a tower, it's a good idea to get in touch with the tower manufacturer or his

Fig. 5-2. Multielement arrays of antennas for low band channels are quite large and heavy. This four-bay array occupies the entire space available and presents a substantial load to the tower.

local sales representative. Towers are tricky things to figure load limitations and you should consult the people who know. For simple single antennas atop a tower, you can find all you need to know in most suppliers' catalogs. If you start to consider mounting a half dozen different antennas all over the tower with some sticking out the side, that's when you need professional help.

Under such circumstances you will most likely be spending quite a bit of money for the tower. To assure proper strength against wind and ice loading, submit your antenna loading requirements to the tower manufacturer and have him propose the best tower for the job. He can help you out with locating antennas on the tower so that they will put the least amount of stress on the tower. The information that the tower manufacturer needs is the weight of the antenna arrays and the wind loading of the antennas. This is usually available as the thrust in pounds projected for the antenna in an 85 mph wind.

LEAD-IN CABLES

Installing lead-in cables involves little more than good mechanical mounting practices after the right cable for the job has been selected. Remember that the lead-in cables are exposed to the full rigors of the elements. Also don't overlook the fact that in vertical runs, the cable must support its own weight from the attachment point.

To cope with water and sun, the cable should be either vinyl jacketed or noncontaminating PVC jacketed. Plain poly jacketing will become brittle and crack when exposed to sunlight thus letting in water to destroy the cable in, most likely, less than a year. Protection from damage from wind vibration is easily achieved by supporting the cables at regular intervals along the lead-in route. Usually a tie off to the most at 2' intervals will suffice. This also eliminates any worry of the cable stretching from its own weight in long unsupported runs.

Horizontal runs of cable should be done with a support wire commonly called a **messenger cable**. Do not lay the cable on the roof where it can be in contact with water for extended periods of time. String a piece of 1/8" galvanized guy wire between the mast and wherever the point of entry is located.

Attach the cables to the end of this messenger wire with a few wraps of a good vinyl electrical tape. Wrap the coax around the messenger wire to form a spiral of about one turn every 2 ft as the cables go along the messenger. This way the messenger cable is carrying the weight of the coax.

If you should find it necessary to use some of the larger cables, such as an aluminum shielded cable, you won't be able to wrap it around the messenger. With large cables or a heavy bundle of cables in horizontal runs of over about 20 ft, you will have to use larger size messenger cable. Guy wire comes in all sizes and for roof top use, 3/16 in. diameter should be more than adequate. Very long runs of 100 ft or more could require quarter-inch messenger but this is seldom encountered in MATV work. This is the type of messenger that is commonly used to support cable along pole lines.

The coax cables are secured to the messenger cable by what is called a **lashing wire**. This is usually a 0.045 in. diameter stainless steel wire that is used to wrap the coax to the messenger. To see an example of how this is done, take a good look at an overhead telephone line that is strung between poles along the street. Wrapping of the lashing wire in these instances is done by machine. For the incidental use in MATV you should do it by hand even if it doesn't come out looking quite professional.

HEAD END INSTALLATIONS

Antennas, wall outlets, and head end equipment are the only parts of an MATV system that show. The head end gear is really the heart of the system and deserves a place of prominence. This equipment is also the most vulnerable part of the system in that it's the location of all the adjustments that make the system play. For this reason alone, all the head end equipment should be located in one cabinet that can be closed and locked after system balance adjustments have been made. The head end cabinet shown in Fig. 5-3 is illustrative of a typical installation. This cabinet has a removable locking cover, not shown, that provides full access to all head end items for adjustment or maintenance as required.

As shown, the equipment mounts well in any standard 19 in. rack. Most manufacturers configure their major MATV

Fig. 5-3. A typical MATV broadband head end equipment installation. Note layout of equipment that routes signals from top to bottom for easy setup and maintenance.

equipment items for 19 in. rack mounting. Smaller items are mounted on specially prepared panels that have predrilled holes in the proper places. The equipment is usually located in the rack so the signals from the antennas enter the top of the cabinet and proceed down through successive devices to come out the bottom. This logical sequence makes for much neater intercabling and ease of following signals during system adjustments or maintenance.

In the vast majority of MATV jobs, the head end is the only place where it is necessary to worry about getting ac power. This is something that you may want to do yourself or something that could easily be done by an electrician. Usually the current load of modern transistorized equipment is quite small. In broadband systems, loads will seldom exceed 100 watts. In strip systems with some closed circuit included, the power required will be well within 250 watts. Only when the system involves cable-powered line amplifiers will the total power get over 500 watts if fully loaded.

This small power load may tempt you to bring power from the nearest convenience outlet to your cabinet. I've even seen some installations where the power was brought in by an extension cord. **This is definitely the wrong way to do the job!** Have a separate circuit run from the nearest fuse box, directly to the MATV head end equipment cabinet. This circuit must be a full three-wire circuit with wire size sufficient to safely carry the full current of the fuse or circuit breaker. The head end cabinet should be equipped with a three-wire plug-mold strip with sufficient outlets to serve all equipment items plus at least one extra outlet should someone wish to use a soldering iron sometime. The three-wire service includes a ground. Most MATV equipment comes with three-wire grounding plugs. It is highly recommended that you use them because they are a part of the equipment that has been tested by the Underwriters Laboratory, UL, for safety from shock or fire hazard.

The UL label on any item of MATV equipment that gets plugged into 115V ac power, is a highly desirable thing. That UL label can be quite important to you. First of all, nothing is UL approved. The Underwriters Laboratories do **not** test the equipment to see if it meets its specs or not. The tests that the equipment **must** pass pertain only to shock hazard and fire safety. After a product has been fully tested and passed, the manufacturer is permitted to add the UL label to his product to show that it does meet these strict requirements. This product is then listed in UL publications as having passed their tests. To assure continued compliance with the safety requirements of UL, the product is retested at various intervals.

Rare as it may seem, you might find yourself facing a situation where a building has burned down and the fire inspector is looking at the MATV installation as a possible cause. In a situation like that, the UL listing of all powered equipment will protect your insurance and you.

THE INDISPENSABLE FIELD STRENGTH METER

Many references have been made throughout this book to the measurement of signal level. You should be aware by now that it is virtually impossible to do any serious MATV systems work without a means of measuring signal strength. The Field

Strength Meter (FSM) is as much a basic tool in MATV systems work as is the VOM in equipment repair. The fact that some installers don't own a FSM and continue to stay in business, is indicative only of the forgiving design of modern equipment. That situation can, however, last only until competition appears with the proper tools and the knowhow to use them.

As we have seen, there is a wide variety of equipment needed to meet the diversified needs of this business. In meters too, there is a variety available commensurate with the needs. For the TV dealer who specializes in antenna installation there are special lightweight meters. These are usually fitted with 300-ohm terminals or provided with a means of easy connection to 300-ohm twin lead. Meters designed for this use are inexpensive but more than adequate for the orientation of antennas and testing of home antenna installations. MATV and CATV work demand a meter with more capacity than available in the antenna installer's meter. Higher accuracy of signal level reading plus the ability to read signal levels of a wider range are among the more important characteristics. The meter in Fig. 5-4, typical of the type used

Fig. 5-4. Typical field strength meter vital to the successful operation of an MATV business. Model FSM-2, courtesy Blonder-Tongue.

in MATV, is fully portable. It's powered by internal batteries. The meter tunes the full range of TV frequencies plus FM and other nearby frequencies for hunting interferences. Readings are directly in microvolts or dBmV over the range of minus 30 to plus 60 dBmV. For those few times when it is necessary to read levels above 60 dBmV, the meter range is easily extended by the use of an external attenuator pad.

Most all meters have some form of detected signal output. Some are labeled "Video Output" while others are labeled "Earphone." This output is very valuable when you're hunting down interferences or making measurements of system performance. It can feed an earphone to help you identify an interfering signal or it can feed an oscilloscope to look at the percentage of ac hum that might appear on signals after passing through an amplifier with a bad power supply filter. The following lists the important tests that can be made with an FSM.

Signal level—Measure level of signals at inputs, outputs, etc. Adjust amplifiers to proper operating level.

Signal balance—Measure the operating level of all channels and adjust each one to be equal (balanced).

Level ratios—Measure the ratio between signals, e.g., picture to sound carrier ratio.

SNR—Measure signal level and record. Remove signal and measure remaining noise. Add 4 dB correction factor to noise reading and compare with signal reading.

Hunt interferences—Use its wide tuning range to hunt out unwanted signals.

Tune traps—Tune meter to unwanted signal and adjust trap tuning for minimum reading.

Tune strip amp agc—See "Balancing the Head End."

Orient antennas—Meter permits you to orient all antennas for maximum reception of signal.

Measure gain—The difference between the input and output of an amplifier is its gain.

Measure loss—The difference between the input and output of a splitter, run of cable, or other lossy device, is the loss through that device.

Obviously the FSM is very important to the MATV engineer and technician. Many other tests involving the FSM with other items of test equipment are used in large systems. Such measurements as for N.F. or cross-mod involve the use of an FSM but are beyond the scope of this book.

BALANCING THE HEAD END

Proper adjustment of the head end equipment in a new system must be accomplished in a careful and deliberate manner. It is important that all equipment is operated within its rated range of signal handling capability. The first check should be made on the antenna signals. Antennas should be oriented for maximum signal and the picture and sound carrier levels of each channel should be recorded on the system layout drawing. It's a good idea to have a variety of attenuator pads available to reduce excessively high levels. As you will find, most suppliers do not specify amplifier input levels and you will have to determine this for yourself by subtracting the amplifier gain from the rated output. This will put you in the right ballpark.

If the master amplifier is a broadband unit, all signal level balancing must be done with the individual antenna inputs. As a general rule of thumb, it will be necessary to adjust the levels of all channels such that there is no more than 6 dB difference between the strongest and weakest channels. If the system is designed to carry adjacent channels, the maximum difference between any two adjacent channels should not exceed 3 dB. The amplifier output must be checked and adjusted to be at the level required by the layout. This is usually within 2 or 3 dB of the maximum output capability of the amplifier type used. Use the gain control of the amplifier to make the final adjustment and do not worry if you find the gain control set near its minimum range. This is not an indication that the amplifier is about to go into overload due to input signal. Input latitude on well designed amplifiers is much wider than output latitude.

Strip amp head ends are much easier to balance than broadbands since you have individual control of each channel. This type of head end equipment permits balancing of signal levels to less than 1 dB between channels. Again it is necessary to check the input signal level to each strip to be sure that the signal is not too strong for that type amplifier. Most strip amps have screwdriver-type gain controls although some are adjusted by means of plug-in attenuator pads.

Strip amps are of two types, **manual gain control** and **automatic gain control** (agc). The gain, or **level setting control** as it is sometimes called, is used to adjust each channel to its rated output. In agc units, it is wise to check the tuning of the agc circuit. Follow the instruction sheet procedure, which is generally:

1. Deliver a signal to the input that is sufficiently strong to make the amplifier reduce its own gain automatically. I recommend about 10 dB more than minimum required to reach full output capability i.e., max. output 66 dBmV, gain 50 dB, minimum input would be 16 dBmV. You should provide an input level of about 26 dBmV.

2. Locate the agc tuning adjustment similar to that shown on the strip amp in Fig. 5-5. Refer to the instructions packed with each unit.

3. Adjust the output level setting control to about mid-range or approximately halfway down.

4. Connect an FSM to the output and tune it to read the level of the picture carrier. At this point the exact level reading is not important.

5. Slowly adjust the agc tuning control for a **minimum** reading on the meter. This assures maximum efficiency of the agc circuit.

6. Adjust the output level setting control for the desired level of operation.

This procedure should be followed for each channel in the system. Do not assume that just because this is new equipment that the agc will be properly tuned. This circuit is very high Q and sharply tuned. Misalignment can occur due to the vibration of shipment from the factory.

As explained in Chapter 4, attention must be paid to the level of lower adjacent sound carriers in adjacent channel systems. The head end design will call for sound control traps on those channels where needed. These traps must be tuned on the job. Check the picture to sound carrier ratio with the FSM. If it is found to be less than 12 dB, insert the sound carrier trap into the appropriate line. Tune the FSM to the sound carrier

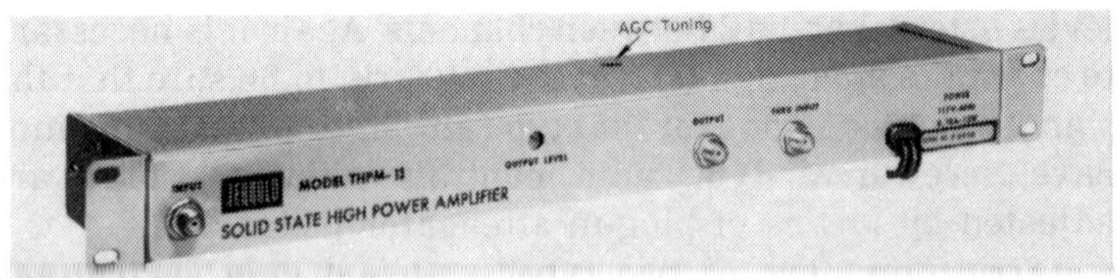

Fig. 5-5. Single-channel strip amplifier. Note location of agc tuning adjustment. See text for procedure. Model THPM-13, courtesy Jerrold Electronics.

frequency and slowly adjust the trap for a minimum reading on the meter. When this is completed, the sound carrier should be approximately 15 dB below the picture carrier level **but not more than 20 dB down.**

All TV signal levels out of the head end must be adjusted to the balanced condition. If the system also carries FM it will be necessary to adjust the FM amplifier to the proper level of operation. This is done first by checking the general level of the FM signals in the band. If they are found to be fairly uniform, plus or minus 4 or 5 dB, then it is safe to set the amplifier so the general level is about 10 dB below that of the TV channels. If the FM station levels are nonuniform, it will be necessary to consider traps for the one or two strongest stations. Adjust the traps for about 20 dB of attenuation of the strong stations then set the level into the system at about 10 dB below the TV signals. This will provide more than sufficient level to any FM receivers connected to the system without fear of overload or limiting the reception to only those few strong local FM stations.

After the head end equipment has been completely balanced it is important to record all operating levels, right on the system layout print. Also record the tuning of all traps. If you have found it necessary to include any attenuator pads that were not designed into the system, mark up the print accordingly. When the job is complete this marked-up print should go back to the shop where a set of "as built" drawings should be made. The system owner deserves a set of these as-builts and it will serve you well to keep a set in your file for future reference. Who knows, you may be called in future to service this system and you can't depend upon your memory or on the owner to find his set of prints. The record of antenna signal strengths in the area is also quite valuable should you get a shot at another job in that part of town.

TESTING ALL LINES

The testing of all feeder lines requires two separate tests. The first test is to determine that each line is properly installed and connected without shorts or opens. Also important is to determine if each line is properly terminated. These tests can be made with an ohmmeter.

Before starting the test, it is a good idea to find out the dc characteristics of the type of taps and splitters used in the system. Test one of each type by measuring the dc resistance between center conductor and ground. If you find this to be an open circuit, as is the usual case, test for continuity between the input and output center conductors. In some cases you will find that certain model taps give a reading of a few hundred ohms between center conductor and ground but almost all units will have continuity between input and output. Splitters and directional couplers are often shorted to ground on an ohmmeter test and this can be considered as normal.

Usually, all feeders originate at the head end location. Each line contains a number of taps and a terminator. If the taps are open circuit to ground, a simple resistance check between center conductor and shield on each feeder line should show one of three conditions:

1. The line is properly installed and terminated because the resistance reading is approximately 100 ohms which represents the resistance of the 75-ohm terminator plus the resistance of the feeder cable.

2. The line is shorted out due to some installation error. This short must be found and corrected before that feeder can be expected to work properly.

3. The line is open which indicates that the feeder is either improperly installed or that the last tap was not terminated. Again this fault must be found and corrected.

If the line contains taps with a dc path to ground, the resistance test will be of little use. Lines with splitters can be tested from the splitter location. In testing from the head end out to the splitter the meter should read some low resistance close to zero. In these cases it may be better to energize the system with signal and go directly to the second test.

The second test is to determine that signals from all channels are getting to all outlets, at or above the design levels. As an absolute minimum you should test for signal level on all channels at the last tap on each feeder. It is also a very good idea to spot check two or three more taps on each feeder just to assure yourself that all is well. A good sampling might be the last tap of each isolation value on each feeder.

CHECK PICTURE QUALITY

Good pictures on the TV screen is what you're striving for as the product of any MATV system. No system can be considered complete until the quality of the pictures at each outlet is tested. Testing every outlet is the only sure way to know if everything has been done right. Depending upon your faith in the people who installed the taps, you may wish to forego 100 percent testing and only look at pictures on those outlets that you sample for proper signal level. This is usually sufficient provided the lines pass all previously mentioned tests.

Picture testing may not be considered scientific but it is the thing by which your system will be judged. For this reason it is essential that you look for all possible problems that could cause a complaint. The basic things to look for are:

1. **Cross-mod**—windshield wiper effect in the pictures.
2. **Snowy pictures**—usually associated with low signal level.
3. **Beats on one or more channels**—usually traceable to FM overload, troubles with converters, or improper sound carrier levels in adjacent channel systems.
4. **Ghosting**—trailing ghosts can be caused by improperly terminated lines. Leading ghosts are caused by direct pickup of signal at the receiver. See "Overcoming Direct Pickup."
5. **General quality**—such as crispness of picture, solid color performance, noise-free sound, or no hum bars in the picture.

The acid test and the most convincing to your customer is a comparison of pictures delivered throughout the system with those directly off the antenna on the roof. This is especially valuable when one of the channels may be questionable and all efforts to correct it have been exhausted. Hopefully you will have known about this situation prior to taking the job and will have already alerted your customer to this possibility.

OVERCOMING DIRECT PICKUP

Direct pickup which produces leading edge ghosts is a familiar problem in MATV systems. The classic approach to solving this problem has been to design the system to deliver high enough signal at all taps to override it. With TV sets that use vacuum tube tuners this is an effective solution. The advent of the transistorized tuner brought a rather severe limit to the maximum amount of signal that can be delivered to a set. This limit, approximately 15 dBmV, is very often not enough to do any good.

In a high percentage of the cases, direct pickup can be conquered by using a 75- to 300-ohm matching transformer that sports a 300-ohm balance of 40 dB or better. Balance on these transformers relates directly to the rejection of unwanted signals that may be picked up on the outside of the coax shield. The cause of most direct pickup problems is the fact that the coax shield is acting like a long wire antenna strung throughout the building. These signals are prevented from getting into the coax by virtue of the shielding efficiency of the coax structure. When they reach the end of the cable at the back of the set they can easily jump past a poorly balanced transformer and appear across the 300-ohm antenna terminals. Well balanced transformers reject these unwanted braid currents by the dB of balance built into the unit. In Fig. 5-6 is a simple bench test that you can make on any transformer to determine its degree of balance. The only piece of test equipment needed is an FSM.

In Fig. 5-6A, you see the initial test setup. Using a clip lead, such as a test lead from a VOM, as an antenna, tune the

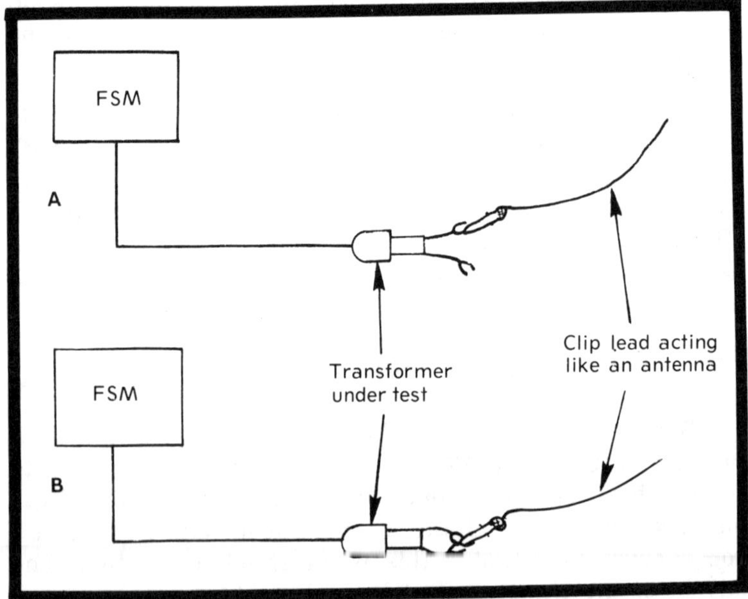

Fig. 5-6. Test bench method of measuring the balance of a 300-ohm to 75-ohm matching transformer. See text for procedure.

FSM until you get a strong signal reading from one of the local TV or FM stations. Record this reading. Be careful not to disturb the setup or move around too much as this will cause a variation in the readings. Next, short the 300-ohm leads together as in Fig. 5-6B and take another signal level reading of the same signal. It should be much lower than the first reading. The difference in dB between the two readings is a good indication of the transformer's balance. Any unit that gives a difference of 35 dB or more can be considered as a good transformer for fighting direct pickup problems.

Severe cases of direct pickup can occur at locations close to the transmitters. Here, even the best balance available in a transformer may not cure the problem. The only way to win in these cases is to convert the channels at the head end and distribute them on the unused channels in the area. Most MATV manufacturers carry a line of VHF-to-VHF channel converters for use in these difficult cases. Follow the manufacturers' recommendations and also be prepared to run into beat interference problems that will require a number of traps to fix. Fortunately this problem does not occur on too many jobs and the need for conversion is rare. For this reason we will not devote any more discussion to the subject.

6 Closed Circuit TV and Other Added Features

Closed circuit or locally originated TV information is often added to MATV distribution systems. The need for closed circuit is most prominent in school distribution systems. Many schools keep a library of video taped educational programs to supplement classroom work. Closed circuit TV (CCTV) also plays an important role in security which ranges from door surveillance in an apartment building to platform monitoring of all stations of an automated commuter rail line.

VIDEO VS RF

Video is the term used to describe the band of frequencies produced by a TV camera. Another term encountered is **baseband** which means the band of frequencies used to modulate a carrier. (Baseband is a more generalized term since it could be any base information such as audio, teletype, video, etc. We'll use **video**.)

The video baseband as encountered in MATV work is not a simple well defined band of frequencies, except as it applies to a standard TV broadcast channel. The MATV engineer,

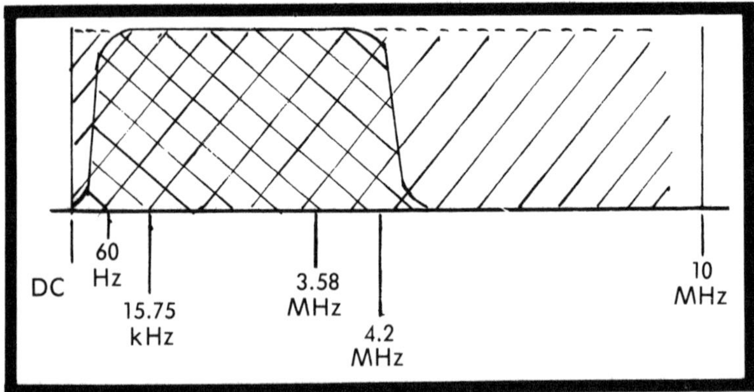

Fig. 6-1. Frequency spectrum from dc to 10 MHz showing video band of standard TV channel, crosshatched area at left, and unlimited band for use in high resolution video applications. See text for bandwidth vs resolution.

however, may find it necessary to provide what is termed **high resolution video.** This involves unspecified bandwidths greater than that of a standard broadcast channel. The video spectrum is shown in Fig. 6-1. The bandwidth for a standard TV channel extends from 60 Hz to 4.2 MHz as shown by the crosshatched area.

Most high quality video processing equipment operates over an extended frequency range greater than that required for TV broadcast. Some equipment is dc coupled or has its range extended down to dc as indicated in the slashed line area. Range extension above the 4.2 MHz limit of the standard TV channel is generally contained within 10 MHz, most running 6 to 8 MHz. The reason for this extra bandwidth is very practical in that it assures the very best signal quality possible over the 60 Hz to 4.2 MHz region.

The detail visible in a video picture with 6 to 8 MHz of bandwidth is considerably more than when limited to 4.2 MHz. The amount of detail visible is directly related to the video bandwidth and is defined as **horizontal lines of resolution.** Horizontal resolution is defined as being the number of black-to-white transitions that can be distinguished along any horizontal line before they all run together into indistinguishable gray. Video test patterns include a resolution test. The vertical wedge shaped lines are calibrated in "lines of resolution" which are no longer separable beyond the bandwidth limits of the system.

Video bandwidth and lines of resolution are predictable by the formula, lines equals 80 times the bandwidth in megahertz. For a standard 4.2 MHz channel, resolution is limited to approximately 336 lines. If 8 MHz of video is viewed, you can expect approximately 640 lines of resolution.

All this raises questions about applying closed circuit to an MATV system. As you know, coax can easily carry the video frequencies, and sometimes they should be handled as high resolution video. Consider the case of a biology teacher wishing to show microscope slides to a class. The students can all file past the microscope, each taking a 10 sec peek or they can all watch a high resolution monitor view of the same slide. Remote classroom activity could be easily transmitted via the cable at video frequencies to a centrally located video tape recorder for later playback.

The advantages of video are many but so are the limitations. It is necessary for the MATV engineer to become familiar with both so that he may use video where it is most advantageous and avoid it in applications where it's not suited.

Advantages of Video

1. Low frequency means low loss in cable and can be carried up to 1000 ft without noticeable distortion.
2. Low frequency means that it can be applied to the same cable carrying rf channels, even to going in the opposite direction.
3. Video is not limited to a predetermined channel bandwidth and can be used in high resolution requirements.
4. Video may be recorded on tape.

Disadvantages of Video

1. Rf distribution system equipment, amplifiers, taps, and splitters will not handle this band of frequencies.
2. Video can be interfered with by 60 Hz power line currents conducted along the coax shield between unequal power line ground potentials.
3. Only one video channel can be carried by one cable.
4. Audio cannot accompany the video on the same cable since they occupy overlapping frequencies.

APPLICATIONS OF VIDEO

Video may be applied to almost any MATV system carrying rf channels. Typical applications call for video to originate somewhere along one of the distribution legs and be brought back to the head end. An example of this is the case where the front door of an apartment house is observed by a TV camera and the picture is fed to all tenants. Special systems equipment such as Jerrold Electronics' J-Jacks, is made especially for schools, thus permitting classroom origination of video programs.

The device that permits the intermixing of video and rf is a low frequency crossover filter. See Fig. 6-2 which illustrates the action of such a filter. Video and rf may be combined onto the common terminal or separated from the common terminal. How this device will permit the use of a nearby riser in an apartment system to carry video back to the head end is

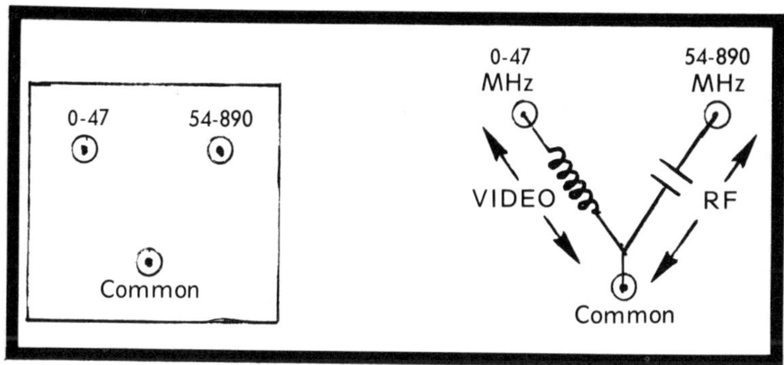

Fig. 6-2. Illustration and simplified diagram for low frequency crossover filter used to mix or separate video and rf.

shown in Fig. 6-3. The video signal is essentially unaffected by the rf taps in the line since the tap side is usually blocked to low frequencies.

In-line passive devices such as two- and four-way splitters will block the passage of video. Crossover filters may be used to get video around such obstacles as those shown in Fig. 6-4. Note that two video systems are in operation on different legs of this system example. This can be done without mutual interference because of the blocking or isolating effects of the rf splitters.

CLOSED CIRCUIT ON RF

MATV systems are not limited to the carriage of off-air signals. Remote origination of video, say in a school

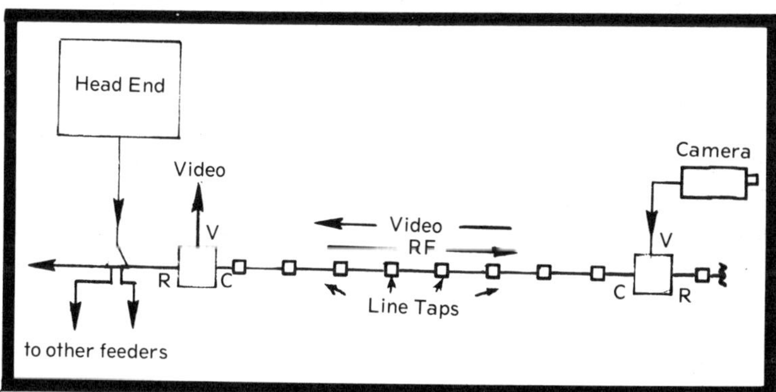

Fig. 6-3. Partial system diagram showing the application of low frequency crossover filters to distribution lines to carry video signal along with rf channels, even in the opposite direction.

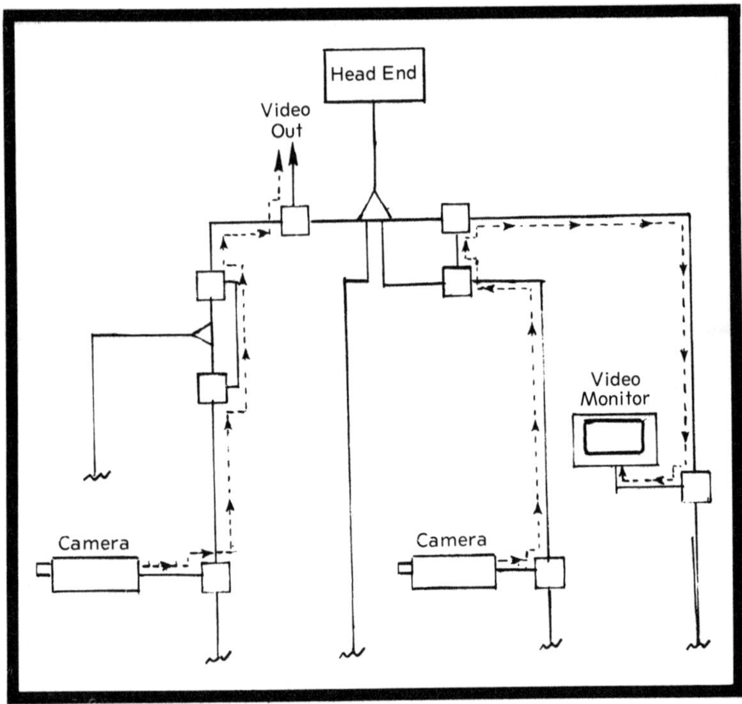

Fig. 6-4. Possible uses of low frequency crossover filter, e.g., bypassing rf splitters, bridging feeder lines, permitting two or more video subsystems to operate without interference.

classroom, may be shown simultaneously to all outlets on one of the unused TV channels. Video returned to the head end may enter a TV modulator for live broadcast throughout the system. Thus an auditorium activity may be seen in all classrooms.

Modulators give the MATV system the possibility of carrying as many closed circuit channels as there are unused channels on the dial. Present day modulators are available for Ch. 2 through 13 and may be intermixed with off-air signals. To provide the maximum number of closed circuit channels in a system, it is wise to consider on-channel UHF distribution for any available UHF air channels. UHF to VHF conversion will only reduce the number of channels available for this purpose.

Video inputs to these closed circuit modulators need not be limited to live cameras. Video tape recorders and film chain cameras are by far the most often used program source. Two different head end diagrams which include the use of

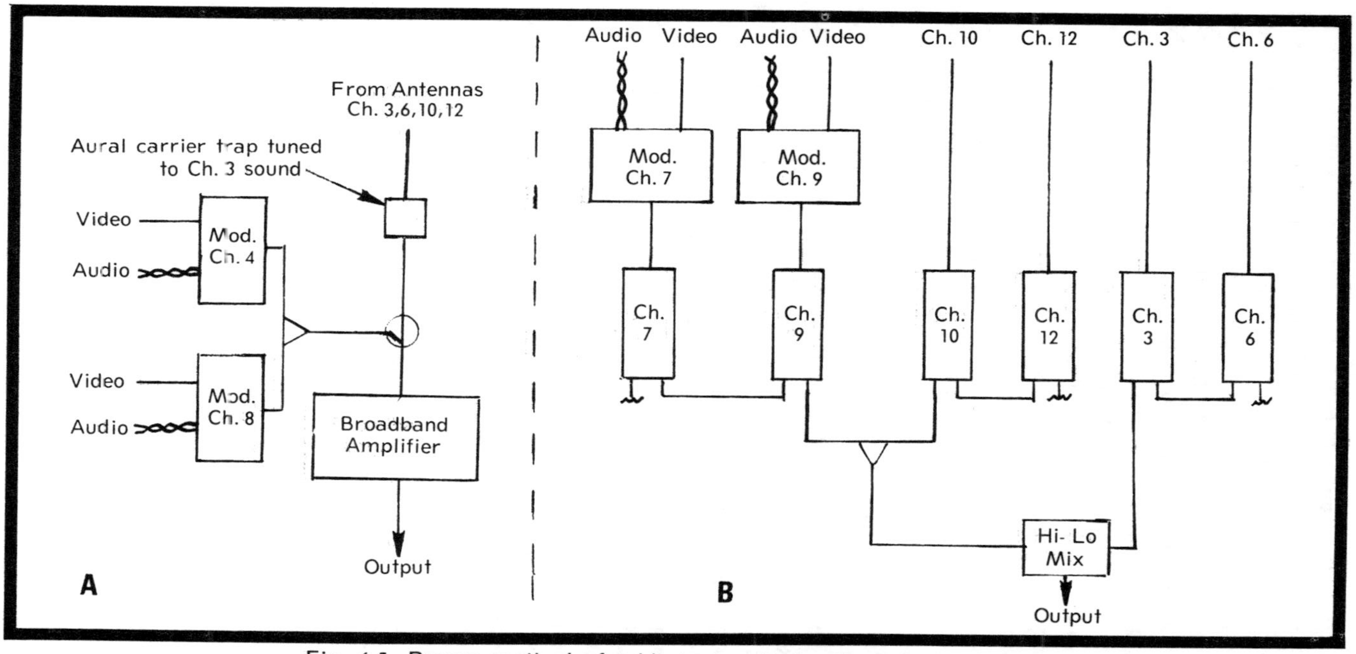

Fig. 6-5. Proper method of adding closed circuit modulators to (A) Broadband head end and (B) Strip amplifier head end. Note use of aural carrier trap to eliminate adjacent channel interference with closed circuit Channel 4.

modulators are shown in Fig. 6-5. In, A, the modulator outputs are fed into the input of the head end amplifier as a means of balancing these signal levels with those received off the air. Note that in selecting Ch. 4 for closed circuit, it is necessary to reduce the level of lower adjacent Ch. 3 sound carrier. In diagram B, closed circuit Ch. 9 is adjacent to air Ch. 10. No sound carrier reducing trap is needed, however, since the modulator-generated sound carrier is controllable or preset at 15 dB below the video carrier level.

Using rf channels for closed circuit gives the designer much more freedom of application. Rf also eliminates any worry of ac hum interference due to ground loop currents. Most of all, rf permits the operation of many channels at the same time, and only standard TV receivers are needed to view all channels.

This ability extends to remote origination of programming leading to two-way operation. Again low frequency crossover networks are used to divide the frequency spectrum into two segments for separate application. The 0 to 47 MHz region is commonly called the **subchannel band**. Up to seven rf channels can be transmitted in this band. Some of the applications of subchannels in two-way systems are shown in Fig. 6-6. Programs may be originated at any tap throughout the distribution system.

Subchannel frequencies fed back into a tap see something less than the loss at Ch. 2 back to the head end, because cable loss is lower at these frequencies. Levels of operation for the subchannel modulators are still about as high as the regular channels because of the losses through taps and splitters.

Converters at the head end receive the subchannel signals and convert them to unused VHF channels. The T8 and T10 channels are converted to VHF Ch. 7 and 9, respectively. It is recommended that turn-around channels, such as in this example, be fed into single-channel strip amps with agc. The agc action will assure proper levels of operation throughout the system in spite of input level variations, caused by varying location of origin. Note that Subch. T12 is converted to VHF Ch. 5 which in turn feeds a demodulator. This arrangement permits the recording of remotely originated signals for later viewing.

Should real-time distribution of this channel be required, the converter output could be patched directly into the Ch. 5

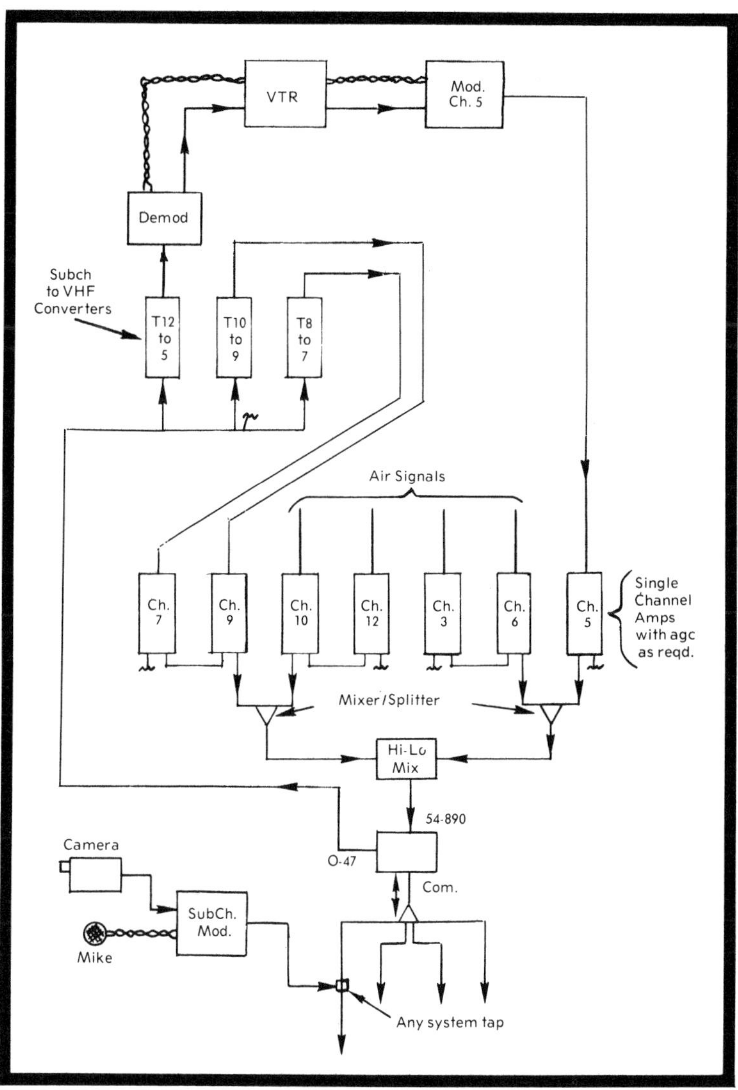

Fig. 6-6. Block diagram of strip head end showing use of subchannels for remote closed circuit pickup from any MATV system outlet location. Agc strip amps at the head end maintain proper signal level into the system regardless of subchannel origination location. CCTV programs may be shown live or recorded for later playback.

strip amp input. It would also be a good idea to include a means of sampling the signal out of the other converters. This would permit connection of the demodulator to these channels for recording purposes.

Educational TV (ETV) system designs vary by the amount of use planned for them. In larger schools, such as a college, there could easily be a demand for all 12 VHF channels to be available for closed circuit. School systems do not have a need to distribute all available entertainment channels. Occasionally there will be programs of academic value broadcast that should be distributed. How this demand can be met is shown in Fig. 6-7. All distribution channels originate in modulators at the head end. Up to 12 channels may be originated from any combination of sources be they local, remote, or off the air. This kind of system requires full time program planning but is successfully used by many institutions with large demands for CCTV.

Modulators

There are a wide variety of modulators available for closed circuit use. A working knowledge of the many requirements of a good modulator is needed to make an intelligent selection. Since there is wide variation in the prices asked for different models it is obvious that the lower priced units have something omitted. The MATV systems designer

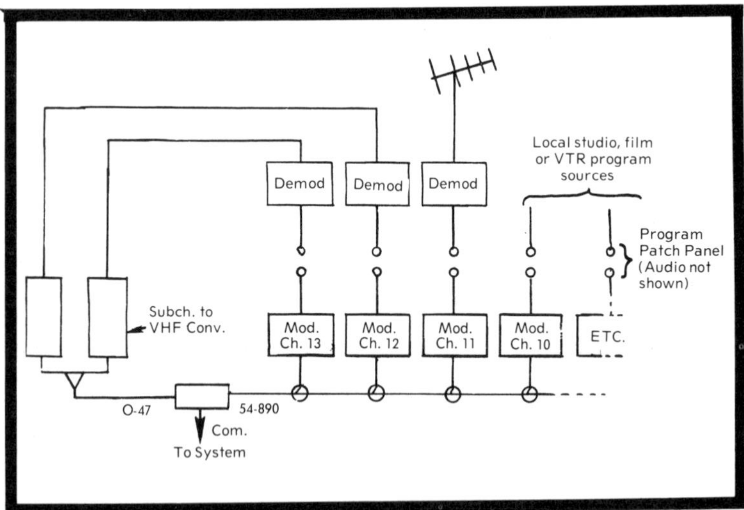

Fig. 6-7. Typical method of designing the head end for a school distribution system where all channels are allocated to ETV use. Demodulator provides access to air channels for distribution or recording of off-air programs of educational value.

must know enough about modulator circuit requirements to know when and where to demand the best or how to make do with less than best.

The following is a discussion of each of the pertinent specs of a modulator and what implications they have to MATV work.

The simplest modulators found are built into self-contained cameras and video tape recorders (VTR). Usually these modulators are tunable over Ch. 5 and 6 or possibly the entire low band, Ch. 2 through 6. This type of modulator will do an excellent job of providing signal to one TV set. Additional sets may be fed from splitters but it is intended that these sets be fed only the closed circuit signals originating from this modulator.

The limitations of this type modulator are many. First, the oscillator is free running and tunable, which means that it is difficult to set the output frequency exactly on channel without expensive test equipment. Second there is nothing to prevent this frequency from drifting with temperature, line voltage, and other common variations. Usually the amount of drift is small, possibly up to 1 MHz, but that's enough to interfere with other channels on the same line. For MATV or other multichannel work, the frequency of operation must be crystal controlled.

Tunable modulators and some inexpensive crystal-controlled modulators are not filtered. This gives rise to other considerations. Since the picture carrier is amplitude modulated, sideband energy is generated both above and below the picture carrier. A vestigial sideband filter is required to eliminate the lower sideband, which can cause interference if used in an adjacent channel system. Do not expect a single-channel strip amp to do the job of a sideband filter. It will help but it won't do the necessary job.

Modulators without filters can be used in nonadjacent system applications but you must guard against other potential difficulties. Without a filter, there is bound to be substantial second and third harmonic output. A Ch. 4 modulator, picture carrier at 67.25 MHz, can also have an output at 134.50 MHz, second harmonic, and 201.75, third harmonic. The third harmonic could easily interfere with high band Ch. 11. Fortunately this interference can be easily eliminated by a simple filter such as one section of an antenna mixing filter or by passing the signals through a single-channel strip amp.

Modulators that also produce sound carriers require special attention. There are two ways of producing the sound carrier. The inexpensive way is to generate it at 4.5 MHz and add it to the video carrier. This produces a sound carrier at 4.5 MHz above **and** below the picture carrier frequency. The one below the picture carrier is extremely difficult to get rid of and is usually ignored. But this can completely wipe out an adjacent channel, so make sure your design won't be compromised.

In high quality modulators the sound carrier is generated at its proper operating frequency or at i-f and converted up to a desired channel. This prevents lower adjacent channel beat interference. It's better, but more expensive.

The sound carrier is FM. It cannot be crystal-controlled and therefore must carry a frequency tolerance spec. Acceptable frequency tolerance requires an automatic frequency control (afc) circuit. Without afc you can get enough sound carrier drift to where sound and picture do not occur together at the same fine tuning of the TV receiver.

Output operating level is another thing to look at. First of all, see if the sound carrier level can be controlled independently of video carrier level. If not, the ratio should be preset with the sound down 15 dB. This will facilitate adjacent channel operation. Next, look to see if adjacent channel operation is recommended at high output levels, above 50 dBmV output. High operating levels can produce unwanted intermod products 4.5 MHz above **and below** the channel of operation. Outboard band pass filters may be required to eliminate these unwanted spurious outputs if internal filtering is not provided.

Aids to operation are features of better quality modulators that can save you money in the long run. Proper setting of video depth of modulation and audio carrier deviation call for expensive test equipment. Better modulators have built in, factory calibrated metering circuits and possibly a front panel meter to indicate proper operation. These features are well worth the small extra price. Other features offered are designed as aids to maintenance, such as plug-in modular construction.

Color Modulators

Color modulators are very demanding of performance specs. Most modulators will accept a color signal input and produce an acceptable quality of picture on a color TV set. If the MATV system is for a college campus or broadcast TV studio monitoring system, you must go all the way. TV broadcasters may use their house system to preview a program to a prospective commercial advertiser. Here the quality cannot be second-rate.

For top quality color performance, the modulator must contain a delay-correcting equalizer. Sound carrier frequency tolerance must be within 1 kHz. Differential phase and differential gain must be within 1 deg and 1 dB respectively. Video response must be flat within 1 dB to at least 4 MHz.

If the foregoing sounds complicated, and expensive, **it is.** The higher quality modulators pay attention to all these details, as this is almost a science unto itself. Suffice it to say that not all modulators are alike. When selecting one for a system, ask the supplier if it will do all the things you expect, including producing a quality picture that will satisfy your customer's needs.

And when you thoroughly understand and apply the knowledge above, and that which is about to be presented, you'll be well on your way to making more money.

PRIVATE OR LIMITED-ACCESS CHANNELS

Modern CATV systems have outgrown the bounds of the 12 VHF channels as allocated by the FCC. The CATV and MATV amplifying equipment, passive devices and cable, must cover the frequency range from 54 to 216 MHz. Within this range of frequencies there is room for more than 12 TV channels. Looking to the future, most manufacturers have extended the frequency range of their equipment to cover out to 300 MHz. This opens the door for even more channel capacity. How these channels are allocated in the cable system is shown in Fig. 6-8.

It can be seen that this arrangement of channels provides a considerable increase in system capacity. First of all, the channel capacity has been increased from 12 to 30. Spectrum space for up to four channels of return information is

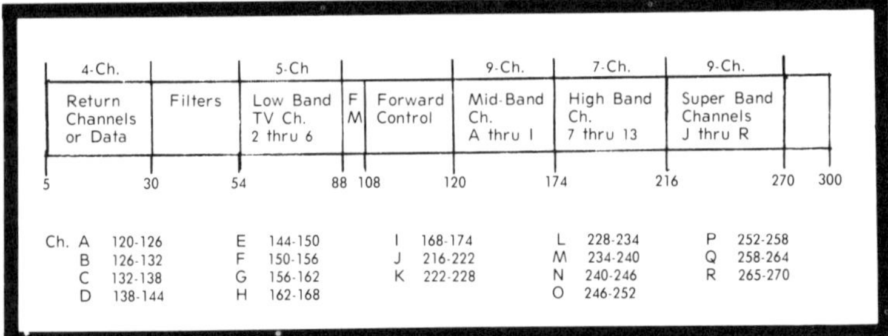

Fig. 6-8. Frequency spectrum from 5 to 300 MHz showing channel assignments for 30-channel two-way CATV systems. Mid-band and super-band channel frequencies as listed.

provided in the 5 to 30 MHz area. It is highly unlikely that this area will be used to carry TV. Most likely this band will be used to carry many channels of command, control, or other applications of digital data. A similar band between 108 and 120 MHz is set aside in the forward direction.

The engineering considerations for the types of systems which carry this density of TV channels is beyond the scope of this book. The application of some of these basic principles to MATV systems, however, opens up some very interesting **extra service-extra income** possibilities.

Modulators are required and available to originate private channels on the nonstandard frequencies. These modulators can feed an MATV system in exactly the same way as with standard VHF channels. The cable and passive devices will handle these frequencies without difficulty. Up to nine private channels can be added below Ch. 7 without any changes in the design approach which is usually figured at Ch. 13 frequency. If more channels are required above Ch. 13, the system must be calculated for that higher frequency of operation.

The beauty of these nonstandard channels is that **no standard TV receiver can tune to them**. Inexpensive, small, top-of-the-set converters are available from many CATV equipment suppliers that will provide access to these channels. An example of where this equipment can be used is in a hospital MATV system. Up to 12 standard TV channels can be carried for normal entertainment viewing by hospital patients. At the same time, private, nonentertainment medical training or observation channels can be viewed by student nurses, interns, medical students, etc., at selected locations which are equipped with special converters. You could make extra money now—such equipment **can be added to existing systems with no changes except the addition of head end modulators.** Set converters operate from typical system tap outlet levels, 0 to 15 dBmV. They function much in the same manner as older, top-of-the-set UHF converters. These converters tune any one of the up to 30 VHF channels. Select an output channel that is not used by any local TV station to avoid direct pickup interference. Most manufacturers offer outputs on at least two of the low band channels.

BACKGROUND MUSIC SYSTEMS

Many hotels and motels have a system for background music wired into each room. By the same token, many do not. In those that did not provide this service, the MATV designer can provide this service as an extra feature to the MATV system.

In very few instances are all 12 VHF channels used on the system. This means that there is unused capacity on most TV dials. It is a simple and inexpensive procedure to add background music over the system on one or more of the unused TV channels. Required for this purpose is a source of program material and a TV channel modulator. For this purpose there are TV channel modulators that have most of

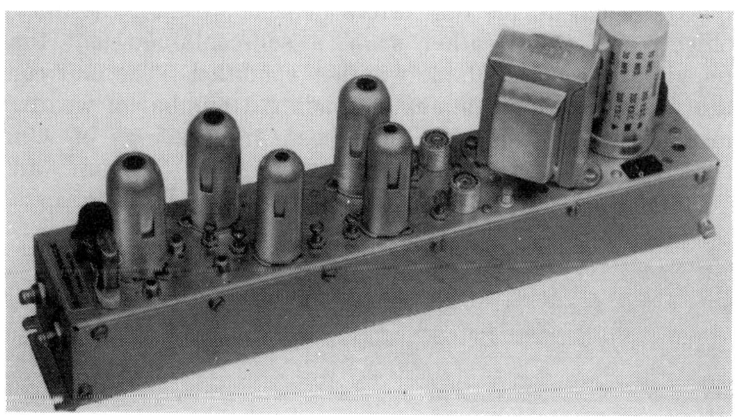

Fig. 6-9. Blonder-Tongue "Audio Master" sound originator for unused TV channels.

the video section removed for the sake of economy. For example, Fig. 6-9 shows the "Audio Master" TV sound originator, by Blonder-Tongue Laboratories.

Program sources are very easy to get by the addition of an AM or FM radio tuner. Program material without commercials is also available from tape as supplied by a number of companies. If the room guest does not want to watch TV, he has the option of listening to background music by dialing one of these channels. In one reported case, a local FM radio station owner used this method of increasing its listening audience. He purchased and supplied the special modulators and a fixed-tuned FM tuner, free to every motel that would permit the installation. The net result was a demonstrable audience increase that commanded higher advertising revenue for the FM station! Rack-mountable AM-FM tuners are available for MATV head ends similar to that shown in Fig. 6-10.

WEATHER SCANS AND OTHER CLOSED CIRCUIT SERVICES

The unused VHF channels on properly designed MATV systems are being looked upon for possible new sources or increases in revenue. Increased revenue can come in many ways. Advertising can be sold and carried on unused channels or features can be provided that will attract more users of all services available.

One such device that offers hope of meeting the above objective is the **weather scan**, a self-contained unit that televises the local outside weather conditions. The unit contains a camera and means of reading a number of weather instruments. Flashed on the screen are readings of wind speed, wind direction, outside temperature, the time, and

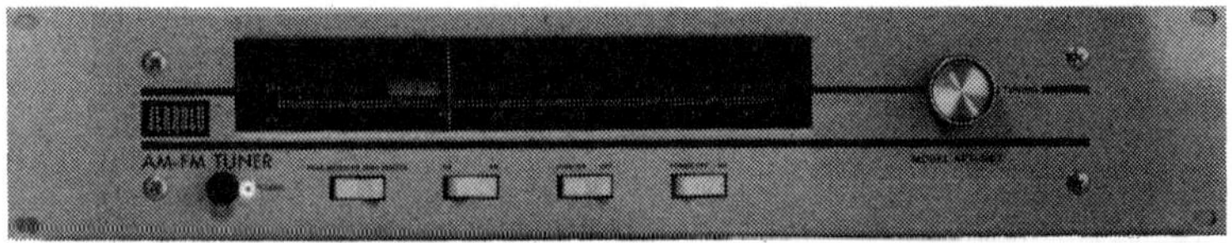

Fig. 6-10. AM-FM tuner for MATV head end. (Courtesy Jerrold Electronics Corp.)

Fig. 6-11. Self-contained weather scan unit complete with camera, weathervane, anemometer, and outside theremometer. (Courtesy Jerrold Electronics Corp.)

possibly other readings. This instant weather information is always of interest and will attract viewers for a quick check of outside conditions.

Also provided in most weather scan units is a means of inserting periodic messages. These messages could be something as simple as tomorrow's forecast or a paid advertisement by the hairdresser in the lobby. To further encourage viewers, you could put the best background music available on the audio that accompanies this channel.

For a typical unit, see Fig. 6-11, a photo of the model TMW-5 weather scan unit, courtesy of Jerrold Electronics. This unit is intended for MATV systems in hotels, motels, or apartment houses. A slide projector may be added so professional quality ads can be presented quickly and inexpensively.

7 Maintenance and Troubleshooting

Not all eventualities are going to be covered by system design. Upon installing a new system you may have to track down interferences or correct problems not anticipated. You may be called to service any system, maybe one of your installations, maybe not. In new or old systems, the problems encountered are dealt with in much the same way. Let's look into the different kinds of reception problems that you might encounter:

Strong signals and no interference. You should always be so lucky!

Strong signals but they contain one or more ghosts. This is probably the most predominant trouble encountered and can be one of the most difficult to eliminate.

GHOSTS

There are two major types of ghosting that can be identified by looking at the TV picture. **Multiple ghosts** that are evenly spaced are most generally caused by mismatches within the system. Signals bouncing between two poorly matched devices will produce such images. This problem can be found by inserting a fixed attenuator, about 6 dB, into various parts of the system until the trouble clears. The fixed attenuator acts to improve the match and may be left in **if you have signal to spare.** The best bet is to replace the items of equipment on each end of that cable run with properly matched new items.

More common is a single strong ghost outlining the picture to the right of the images when looking at the TV screen. This is caused by the reception of two signals, one directly from the transmitter, the other reflected from some object. The reflected signal is weaker than the direct signal and delayed in time due to the longer distance it has traveled in getting to the receiving antenna.

Contrary to popular belief, the object from which such a reflection occurs is seldom the side of a mountain many miles away but rather some object like a building, water tower, bridge, etc., usually nearby. To produce a ghost spaced a quarter inch on a 21-inch TV set requires a time delay equal to only 800 ft. Thus a building 800 ft off to the side or 400 ft behind the antenna could produce a ghost with a quarter inch displacement. By contrast, a mountainside could cause a reflection but if the mountain is 5 miles distant the ghost would be displaced 8 inches.

The graph in Fig. 7-1 can be used to determine the multipath distance for ghost displacements of approx. 1 / 16 to 1 inch on various size TV screens. Ghosts with time delays equal to 230 ft or less cannot be distinguished as separate images. They are limited by the bandwidth of the TV receiver and appear as smears of a softening effect to otherwise sharply defined lines.

Remember that reflected signals can arrive from any angle! The actual location of the object producing the reflection cannot be determined from Fig. 7-1. This is because the object which causes the reflection can be located anywhere that produces a multipath reflection equal to the ghost displacement. The oval shaped line on the inset of Fig. 7-1 illustrates the possible locations of a reflecting object that would produce a ¼ in. ghost on a 21-in. screen. Keep in mind that the further away the reflecting object, the larger it must be to produce a visible ghost. Coupling this knowledge with the fact that most antennas have good rejection to signals from the side says that the most likely place is from the front or back of the antenna. If from the front, it must be a large object like a bridge. If from nearby, it's almost anything; a building, water tower, or other local structure. Most likely it's the latter and you can tell the difference by rotating the antenna.

If the ghost remains unchanged, it's probably from some large object miles away but toward the transmitter. This happens infrequently but in these cases there is really very little that can be done to improve reception. If it's any consolation, you can bet that everyone else nearby is suffering

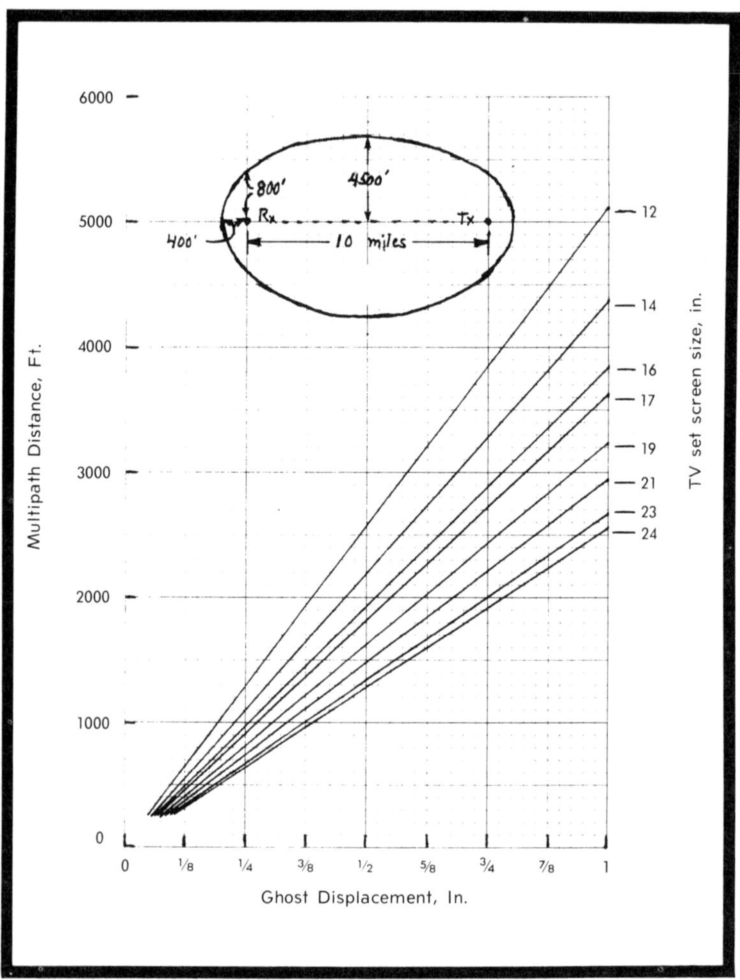

Fig. 7-1. Graph indicates the extra path length required for reflected signals to produce ghost displacements of 1/16 inch to 1 inch on various TV screen sizes.

from the same problem. The only thing that's worth trying is **lowering** the height of the antenna, if practical. This will tend to reduce the amount of total received signal but usually the ghost strength drops more rapidly and a satisfactory height may be found.

In rare cases, if the ghost is very strong, it may be practical to zero the antenna at the ghost itself and build up its signal level to where it is stronger than the primary signal. But this method, of course, requires much luck, and labor, and may be self-defeating. It's an item to keep in mind, however, especially in mountainous fringe reception areas. Here, an FSM is essential.

If the ghost changes appearance with antenna rotation, it's undoubtedly from a nearby object. In these cases, antenna location itself will have a big affect on the ghost strength. Usually it is sufficient to rotate the antenna until clear pictures are received. This will compromise received signal strength but is a small price to pay for clean pictures. Probing the roof for a ghost-free location is also recommended. If ghosting persists and you can identify the source, the horizontally stacked co-channel array described in Chapter 4 can be used for additional rejection of this unwanted signal.

ELECTRICAL INTERFERENCE

Electrical interference is identified by two bands of snow-like noise horizontally across the picture. These bands may be steady with the picture or slowly rolling up or down. If this interference is of an intermittent nature, it's usually caused by the arcing of brushes on an electric motor. The cure for this is to find the offending machine and have it repaired. It may take some time to track it down but the idea is to match up the most likely cause with the pattern of on and off. Fairly regular intervals of on and off, both day and night, could be an air conditioner, a refrigerator, a furnace, etc. A random pattern in the on-off sequence indicates some man operated machine such as an electric drill, sander, or other household appliance such as garbage disposal, electric shaver, etc.

Arcing produces rf interference over the entire frequency spectrum. The lower the frequency, the more intense the interference. That is why it is more pronounced on the low channels. An FSM and a small antenna will help you locate the source. When the interference is not strong enough to make the needle move on the meter, use the earphone to determine the direction toward the strongest signal. After finding the source, tactfully call it to the attention of the machine owner. He undoubtedly doesn't know that his machine is badly in need of repair and is about to fail.

For cases of continuous interference, it's usually power line problems such as cracked insulators, arcing transformers, or poor electrical joints. **A general complaint to the**

power company won't do the job. A high gain antenna and your FSM can tell you the direction toward the source if it's a single defective item. Driving in the direction of the interference with the AM car radio tuned between stations can be used to locate the source. If you are fortunate enough to pinpoint the trouble to within about a block, a call to the power company is in order and corrective action is usually forthcoming.

IGNITION NOISE

Interference generated by automobile ignition is generally noted in weak signal reception areas. This type of interference appears on the screen as randomly spaced horizontal streaks. The problem is insufficient signal to interference ratio. Since there is nothing you can do to decrease the amount of ignition noise generated, your only hope is to increase the amount of received signal. How to use stacked antennas for increased gain is covered in Chapter 4. Two antennas stacked in array will give you almost 3 dB more signal. It will also decrease the interference because of the sharper beam.

You cannot expect to completely eliminate ignition interference. You **can** expect, however, to make noticeable improvements in the degree of interference that are well worth the effort.

BEATS AND WIGGLY LINES

Beat interference is a very severe type of picture quality problem. The presence of beats in a picture can range from a mild form, called **herringbone**, to a complete wipeout. The potential sources of beats are many. Unfortunately the source of any given beat cannot be positively identified by merely looking at the affected picture. Some clues can be gained, however, and we will discuss the more common causes and remedies.

Overload by FM is probably the most often encountered type of beat interference. In broadband systems it can affect a single low-band channel, usually Ch. 6, any single high-band channel, a combination of channels, or all channels. In single-channel strip systems, this type of interference is usually limited to Ch. 5 or 6. The severity of the interference and the number of channels involved is dependent upon the strength and number of FM stations received.

It is first of all necessary to identify the cause of the beat as being FM. This can be done by measuring the strength of received FM channels with your FSM. If one or more FM stations are coming from the antenna at levels exceeding the TV channels, chances are that FM overload is the cause of the beat.

There are two types of traps that will eliminate this problem. Single frequency traps are used to provide a lot of attneuation on a single station permitting all others to go through the system. This type of trap would be used where FM signals are desired on the system. These traps are tunable over the entire band and are properly set with the use of an FSM. Tune the meter to the offending channel and adjust the trap for minimum signal reading or maximum attenuation. Traps with two adjustments should provide about 40 dB of attenuation.

It is strongly recommended that each adjustment of the trap be slightly misaligned from the maximum attenuation setting. This should be done by adjusting one control slightly in the clockwise direction until the offending signal increases 4 to 5 dB. Adjust the other control counterclockwise to increase the signal another 4 to 5 dB. This will leave you with about 30 dB of rejection to the offending station. The object of this **purposeful** misalignment is to assure stability of the trapping effect. Trap tuning will drift slightly with temperature variations and the benefits of your efforts could easily be lost. This point is illustrated in Fig. 7-2A, which shows the initial tuning response while B shows what happens to the trapping effect after slight tuning drift with temperature variations.

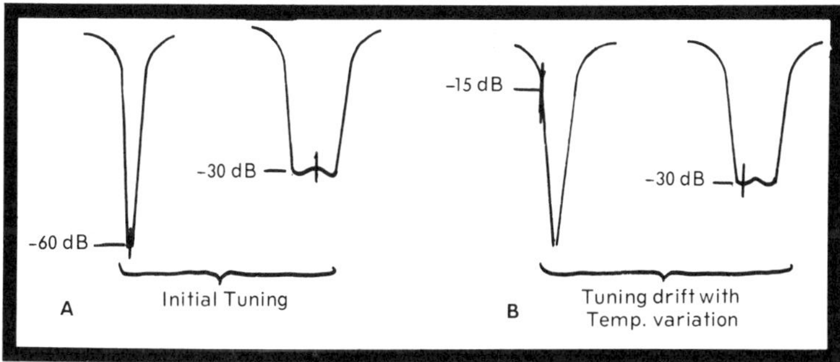

Fig. 7-2. Proper method of tuning interference traps. All high Q traps are subject to drift with variations in temperature. Note effect of drift on trapping efficiency if initial tuning is for maximum.

The other type of trap available for FM overload problems is the **band reject filter**. This device is factory tuned to provide approx. 20 dB of rejection to all channels in the FM band. A filter of this type would be used where FM is not used over the system or where there are a number of offending strong FM stations being received. There is no worry about drift with this type of filter but rejection is limited to approx. 20 dB.

Placement of the trap or **reject filter** in the system is important. If you use a mast mounted preamp on the antenna, don't expect the trap to eliminate the problem by inserting it in the head end rack downstairs. In most cases, the damage is done in the preamp. For proper effectiveness, the trap must be placed between the antenna and the preamp.

INTERMITTENT INTERFERENCES

Some interference is intermittent. The interference can range from a mild beat to the complete wipeout of one or more channels. The fact that the interference is intermittent gives you some information about where to start looking. It's rather obvious that it is not an FM station but is something that's not transmitting all the time.

To cope with this situation, it is necessary to identify the source of the problem. Your FSM is the tool that will do the job. Tune the meter in and around the area of the affected channel during the time that the interference is present. You should be able to find an extra signal. With the earphone connected to the meter you should also be able to hear the signal of the interfering station and identify its source. Most likely it will be some low frequency transmitter such as Citizens Band, at 27 MHz, Amateur Radio at 28 MHz or 50 MHz, or Public Safety such as Police, Fire, etc., at 29 to 50 MHz.

Radio services in these bands are used intermittently and many are mobile. The interference symptom to TV is like FM, the overloading of a preamp or other amplifying equipment by a strong signal. To cure this problem requires a band pass filter that will pass the TV bands and reject anything below 54 MHz. Such filters are available and their prime applications were discussed in Chapter 6. The filter is a crossover network of two filters for splitting the TV band and the subchannel band. In Fig. 7-3, this filter is shown in an antenna downlead for purposes of rejecting all low frequencies below TV (54 MHz).

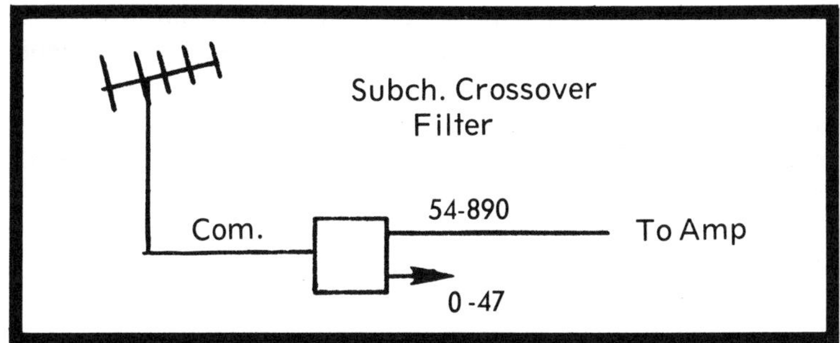

Fig. 7-3. Diagram shows use of subchannel crossover filter in antenna down-lead to reject interference from any low frequency source below 54 MHz.

OTHER WIGGLY LINES

Beats in pictures that are not associated with FM or other interfering signals are all too common. This situation usually arises in systems where UHF is converted to VHF or where modulators are used to provide closed circuit channels. In general, these beats are associated with unbalanced or improper operating levels. Again, your FSM will come to the rescue.

One of the quickest ways of finding the source of these beats is to disable each channel, one at a time. Start with U to V converters then try any modulators in the system. Usually disabling one of them will cause the beat to go away. This channel should then be checked thoroughly for the cause of the beat. The things to look for are:

1. **Balance**—is the level of this channel higher than all others?

2. **Sound carrier level of lower adjacent channel**—is it at least 15 dB below the picture carrier level?

3. **Operating levels**—is the signal level into and out of each device within the specified range?

4. **Something unexpected.** Test the output of the converter or modulator for spurious outputs at frequencies other than expected.

If you find an unwanted output, it could mean that the unit needs to be repaired or replaced. In the case of a converter, it could be because a new station is now on the air and you must add a trap or filter.

Chronic failure to solve these problems is another indication of something wrong. Most of these troubles are connected with **operating level.** If after you do everything right and despite some improvement, you still have trouble, suspect the accuracy of your FSM. Meters are electronic devices and are subject to vibration and rough handling that can cause accuracy to change. If you suspect this, check your meter against another meter or return it to the manufacturer for calibration.

ROLLING PICTURES, HUM, AND SYNC CLIP

Pictures that do not hold sync can be caused by a defective MATV distribution system. Before embarking on the trail of trouble, **make sure that the problem is in the system and not the set!** Check with a couple of neighbors: if they're having trouble also, it's the system; if not advise Mrs. Jones to call her TV repairman.

Ac hum modulation of the MATV signals being distributed is one cause of picture rolling. This type of problem also appears on the screen as one or two horizontal black bars which may be stationary or rolling through the picture. Early stages of hum problems are seen as lightly shaded bands which do the same thing.

One bar or band in the picture indicates 60 Hz hum, two bands indicate 120 Hz hum. This is usually caused by a bad filter capacitor in an amplifier power supply. One band indicates problems in a half wave supply while two bands is trouble in a full wave or bridge rectifier circuit. Trouble in only one channel can be traced to that channel's strip amp or preamp. Trouble on all channels is indicative of a broadband amplifier problem.

Sync clipping is another source of picture rolling problems. This is associated with strip amp systems or other single-channel processing devices such as preamps, converters, or modulators. The problem here is that the unit is being driven so hard that the highest part of the signal, the sync tips, is being clipped off. The most frequent source of this problem is with agc strip amps. What happens is that with time and aging, the tuning of the unit's agc pickoff circuit can shift. With less and less agc action, the amplifier thinks it is seeing a decrease in input signal level and increases its own gain to compensate. This causes the output signal level to increase to the point where the amplifier is overloaded. It doesn't take much with transistorized equipment to produce serious troubles. Three to four dB above rated maximum output for a strip amp will often be enough to produce this kind of complaint.

Your FSM will tell you the story right away. If you find the output level of a strip amp to be high, don't settle for just resetting the gain control! Check the tuning of the agc by following manufacturer's recommended procedure.

Sync clipping can occur in other single-channel devices like preamps and converters. Again the problem is too much signal. Check your recorded levels on the "as built" drawings. The cause could be something as remote as the TV station having increased its power. But chances are that component failure within the unit is the problem and repair or replacement is the answer.

CROSS-MODULATION, THE WINDSHIELD WIPER EFFECT

Cross-mod is the kind of distortion that produces picture degradation in broadband systems. Like most other troubles, it's caused by signal levels being too high for the amplifier to handle. The problem is akin to that of FM overload only here the overload is caused by the TV signals themselves. The effect on the screen is to see vertical bars wiping back and forth across the screen. If the interference is strong enough or sufficiently steady, you can make out the picture of another channel in the background. Don't confuse this with a stationary ghost.

Cross-mod usually occurs in the output stages of a broadband amplifier and is caused by one or more channels exceeding rated output. To cure this problem you must adjust the levels of eacch channel so they are balanced with all other channels and to operate each channel within the amplifier rating. Amplifier components going bad, especially tubes in older equipment, can be the cause of cross-mod. Suspect this when cross-mod still shows after proper level setting of all channels.

Don't be surprised to find the system operating at unusually high levels. This can easily occur through the aid of some helpful tenant or maintenance man who turned the gain full up to getmore signal as the result of the first complaint that didn't go to you!

A FEW WORDS ABOUT SNOW

Snow is noise in the system that is strong enough to see in the picture. Usually it is caused by weak signals from the antenna. Your friend here is not only the trusty FSM but a record of system signal levels. It's one thing to confirm a complaint about snowy pictures with your meter and another to know if there's anything you can do about it.

You might be responding to a complaint of a new tenant in an apartment building who never saw a snowy picture before. You could waste a lot of time trying to fix that one. If system records are kept and you measure the same antenna signal as when the system was first installed you would suspect a bad preamp and make the necessary checks. If you find that signal strength is substantially below the initial readings, look for a bad antenna, a bad lead-in cable, or watch that new skyscraper being built across the street in front of your antenna.

SYSTEM GHOSTS

Problems not associated with active head end equipment usually show up as ghosts in the picture. There are two kinds of ghosts that cause trouble—leading ghosts and trailing ghosts. The ghosting under discussion here assumes clean, ghost-free signals coming from the antenna to the head end.

Leading ghosts are the result of direct pickup. Signals received by the TV set directly from the station arrive at the set slightly ahead of those via the cable. The result is a ghost along the **left** edge of picture images. This problem is fully discussed in Chapter 5 regarding new systems; however, it is not unusual to get a direct pickup complaint from a system that was initially clean.

Most likely this trouble stems from someone having destroyed your initial efforts to defeat direct pickup. Signal input matching transformer leads and sometimes the entire unit will be damaged when people move their furniture around. When people move they sometimes take the transformer with them. The next tenant might use lamp cord to connect his set. Damaged transformers are often discarded when the tenant finds that connecting the coax directly to his set restores operation. These are rather obvious and easy to find things but you would be surprised how often they occur.

More subtle than this is the person who uses the second outlet in his apartment to feed his FM radio. He may run 50 ft of twin lead under the rug from the outlet to the FM tuner. This can become a good long wire antenna to feed direct pickup signals back into the system. Finding this one could be tough and requires visiting each outlet on the offending riser.

Trailing ghosts or ghosting around the **right** hand side of picture images are primarily caused by system reflections. The reason for this to happen as a trouble complaint is usually associated with somebody's tinkering with the system. A tenant may find that his set is getting old and now producing snowy pictures. If he gets into the tap and jumps the isolation components, he will get more signal but introduce a reflection that will bother everyone else. The system owner may take it upon himself to extend the system to one or more outlets and it's an even money bet that he won't terminate the extended line. Also, don't overlook the possibility that the line was never terminated when the system was initially installed. Here an ohmmeter will usually tell the story. Line termination can usually be tested from the head end by simply measuring the resistance between center conductor and ground. An open or shorted line will tell you what to look for.

EQUIPMENT REPAIR

The MATV system equipment is not difficult to repair. The complexity of such equipment is rarely equal to a black and white TV set. The only problems associated with repair of this gear is adequate information from the manufacturer and the necessary test equipment. Hopefully, you'll also have "as built" drawings.

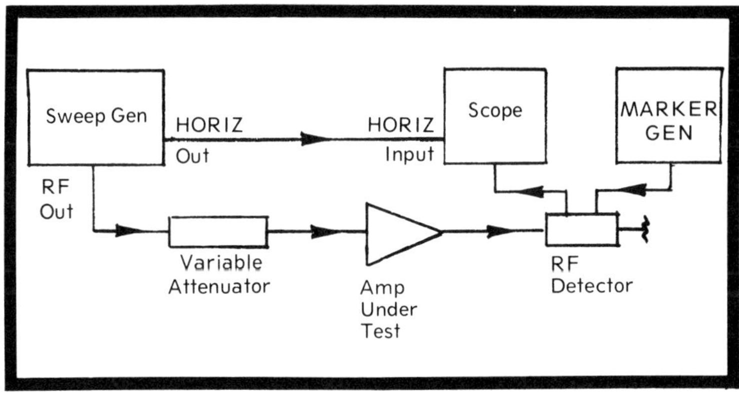

Fig. 7-4. General test equipment setup to perform sweep tests. Sweep may be used to measure gain, loss, and response of amplifiers, filters, etc.

Repair of broadband amplifiers can usually be done in a service shop with a minimum of test equipment. Such repairs are usually quite reliable and full operation is easily restored. This is because broadband equipment has a minimum of frequency-sensitive elements that can be affected by parts replacement. However, some parts, especially transistors, used in MATV amplifiers, are sometimes very specialized. Do not attempt to substitute standard 2N transistors for manufacturer-selected devices! Manufacturers will sell you exact replacement parts at nominal prices. This is the only way to assure full signal-handling capability of the repaired unit.

Single-channel devices should only be repaired if you have a means to check the alignment response. This requires a sweep frequency generator, impedance matched rf detector, calibrated rf marker generator, variable attenuator, and a decent oscilloscope. Fig. 7-4 illustrates the test setup for sweep-testing an amplifier. With some of the more tricky filtering circuits, you can add the need for an impedance bridge.

All this adds up to a rather large expenditure for test equipment. If your service work warrants it, by all means make the investment. If you only have occasional need to do repairs, keep a small stock of amplifiers as loaners and use the repair department of the equipment manufacturer. Repair rates are reasonable, usually time and materials, but the time lost due to shipping is usually a few weeks. The loaner amplifiers will keep your customer happy during the time his unit is out for repair.

Interconnecting With CATV

The advent of FCC regulations on CATV, in early 1972, released a freeze on CATV construction in the major cities throughout the United States. The MATV engineer must be prepared to design a system where the source of signal input may as well be a feed from the local CATV system as the traditional antennas on the roof. This type of system design requires a working knowledge of CATV considerations because any MATV system that distributes signals from a CATV system must function technically as part of that CATV system.

This chapter is not presented as a complete course in CATV engineering but rather as an introduction to those special concerns relative to CATV that will have an effect on the operation of any MATV system you are called upon to design. There are three basic considerations which the MATV designer must think about including, over and above that needed for MATV alone:

1. Most CATV systems carry all 12 VHF channels. Many CATV systems are capable of carrying 21 or 30 channels in the spectrum of 54 to 300 MHz.

2. All CATV systems introduce degrading noise and cross-mod to the signals. The MATV system designer must consider this and limit the amount of MATV-added degradation to reasonable limits.

3. Modern CATV systems are designed for two-way communications; this also places additional burdens on the MATV system design.

CHANNEL CAPACITY

The number of channels carried by an amplifier has a direct effect upon the amount of degradation generated in the form of cross-mod. With three or more channels in the pass band, well performing amplifiers will follow the rule of **3 dB per double**. This boils down to mean simply that if you double the number of channels into an amplifier it is necessary to reduce the maximum output level for each channel by 3 dB.

A review of typical MATV amplifier specifications shows that most manufacturers state their product's output capability at minus 46 dB cross-mod with seven channels input. Since most CATV systems carry at least 12 channels, this can be looked upon as being almost double the catalog rating for channel handling capacity. For this reason alone, any MATV product that is going to be used with CATV signals should be derated by almost 3 dB. I recommend 2.5 dB derating since 12 channels is not quite a double over the 7 channel catalog rating.

When confronted with more than 12 channels it will be necessary for you to consider special equipment. The problem isn't the additional 3 dB derating of output to go from 12 channels to 24, that's easy. The problem is how to cope with the **second-order beats** that are generated in regular MATV amplifiers. These beats fall into the mid-band range, 120 to 174 MHz, and the super-band range, 216 to 300 MHz. The mid- and super-band ranges are of course the bands used to carry these additional channels.

The solution to this is to borrow a page from the CATV catalog and use equipment of the push-pull design. Push-pull equipment is already specified for carrying the large number of channels and will not have to be derated as in the case of MATV equipment. The value of the push-pull design goes back to the first days of high fidelity audio. There, harmonic distortion was canceled out by a push-pull amplifier design. The same design philosophy is applied to this problem with equal success.

In summary then, standard MATV equipment may be derated for use with CATV signals as long as the total number of channels doesn't exceed the standard 12 VHF channels. When it is necessary to amplify and distribute CATV signals with more than 12 channels, it will be necessary to use push-pull CATV equipment or specially designed push-pull MATV equipment that is compatible with the CATV channel density.

LIMITING THE DEGRADATION

Degradation in the form of noise and cross-mod is present in all signals delivered by any amplified system. In MATV work it is permissible to use 100 percent of the allowable degradation at the head end. This is economically sound because it permits the design of large systems while the degradation generated is always kept just below the level of visibility on the TV screen. This is also true in CATV work; however, the degradation generated must be spread over all the amplifiers in the CATV system with each one limited to only a small percentage of the total. What this says is that depending upon where you are located on the CATV system you can expect to receive signals that have from 50 to 85 percent of the permissible degradation already used.

Let us first take a look at noise. As we know, all amplifiers add some noise to the system. The amount of noise added is a function of the N.F. of the equipment used, as was explained in Chapter 1. In CATV systems where the signal must be amplified many times to carry it over the long distances involved, each amplifier along the way adds to the total noise in the system. This fortunately follows the 3 dB per double rule. This is to say that for every double in the number of amplifiers in cascade in the system, the amount of noise in the system is increased by 3 dB.

Large systems get to be 30 to 40 amplifiers long at some of their extremities. This represents about 5 doubles. Let's calculate:

Number of amps in cascade	System noise performance
1	Catalog N.F.
2	Add 3 dB
4	Add 6 dB
8	Add 9 dB
16	Add 12 dB
32	Add 15 dB

Since metropolitan area CATV systems are bound to be large, it will be necessary to use the 15 dB noise derating to be safe, regardless of location. The fact that noise builds up along an amplified system means that the signal delivered to a building distribution system must be strong enough to overcome any noise that the MATV amplifier may add. Minimum signal level permissible can be determined by seeing what it takes to meet a 45 dB SNR. Assume a typical N.F. of 8 dB for the type of amplifiers used in the CATV system. After 32 amplifiers in cascade the noise level will have increased by 15 dB. Thus we can see that minus 59 dBmV basic noise plus 8 dB N.F. for one amplifier plus 15 dB for cascade noise equals minus 36 dBmV of noise at the system extremities. And 45 dB above that gives us a signal level of 9 dBmV as the minimum input to any amplifier. If you follow this minimum you will operate your MATV system in such a way as to assure minimum addition to the total noise in the pictures delivered.

Cross-mod distortion is the more serious problem and the one that requires the most care on your part. Just as noise adds up in a cascaded system, so does cross-mod. Because cross-mod is a repetitive type of distortion as opposed to noise (which is random), it adds together as a **voltage** or **6 dB per double**. What this says to us is that if the signals delivered by the CATV company had a cross-mod level of minus 46 dB and you were to operate the MATV system at minus 46 dB cross-mod, the total cross-mod delivered to the TV sets would be minus 40 dB. As we have stated throughout this book, minus 46 dB cross-mod is all that you can stand without showing windshield wiper in a picture. This makes it obvious that you can no longer operate the head end equipment at its MATV output rating.

Fortunately, operating CATV companies do not use the minus 46 dB cross-mod limit as a standard. The CATV standard for cross-mod is minus 57 dB. Since the two major portions of a CATV system, trunk and feeder, both operate at the minus 57 dB cross-mod level, the worst case situation should find you with a cross-mod level of about minus 51 dB. This as you remember is about the point where the windshield wiper effect of cross-mod just disappears **on a blank screen**. Your objective must be that you do not add substantially to the cross-mod already in the signals. Graph 7 in the appendix shows how noise and cross-mod add together. To repeat: Note that equal levels of noise add together to be 3 dB stronger. Equal levels of cross-mod add together to be **6 dB** stronger.

Using this graph you can see that if you added an equal amount of cross-mod the total would be minus 45 dB. If we were to follow the CATV standard of minus 57 dB, the contribution would only be 3.6 dB for a total of minus 47.6 dB cross-mod. This is the recommended operating situation for MATV systems that are fed from a CATV cable. Noise and cross-mod are only slightly increased but are still within the limits of tolerance.

DERATING MATV FOR CATV SERVICE

As has been discussed, systems with more than 12 channels will require the use of push-pull amplifiers typical of CATV products. In systems that deliver 12 or less channels, MATV equipment may be used if it is properly derated for noise and cross-mod.

The first consideration is **noise**. Here there is nothing to be done with the operation of the amplifier. However, when the MATV engineer makes his selection of the proper amplifier, he must include a consideration for a low N.F. A good guideline here is to limit the N.F. to not more than that of the products used by the CATV company. In any event this should not be higher than 9 to 10 dB. The next consideration is in the amount of signal delivered to the MATV system head end by the CATV company. It is typical for the CATV company to deliver a level of 6 dBmV to its subsrcibers. **This is not enough!** A minimum level of 10 dBmV should be made available to you, if for no other reason than that is the typical level CATV uses as minimum in its own system.

Cross-mod is another story. Here the amount of this distortion generated is directly concerned with the output levels at which you operate the head end amplifier. Typical MATV amplifier specs list the output capability of a broadband amp at minus 46 dB cross-mod with 7 channels. For these amplifiers to be used in a system served by CATV it will be necessary to derate the output for operation with 12 channels at minus 57 dB cross-mod. Using our knowledge of how cross-mod is developed in an amplifier, we can predict exactly how much it must be derated to meet this kind of service.

First it is necessary to derate for the difference between the 7 channel specs and the 12 channel loading. Here the 3 dB per double rule comes into play. Fourteen channels would be one double over the 7 channel rating for the amplifier. Since we are only increasing to 12 channels, the derating factor is only 2.5dB. This means that the amplifier must be operated at 2.5 dB less output per channel than the 7 channel rating. This, however, would still leave us at minus 46 dB cross-mod. To reduce the cross-mod from minus 46 dB to minus 57 dB involves a reduction of 11 dB. Fortunately we get a 2 dB reduction in cross-mod for every 1 dB of output level reduction. This means that the amplifier must be operated an additional 5.5 dB lower in output. The total reduction is therefore 8 dB below MATV rating.

A review of broadband amplifiers available to the MATV market shows a grouping of products at approximately 5 dB intervals from 40 to 60 dBmV output capability. Using this as a guideline the following table compares the 7-channel MATV rated amplifiers to their 12-channel CATV handling capacity.

Example Broadband Amp	MATV rating 7 Ch. at minus 46 dB cross-mod	CATV rating 12 Ch. at minus 57 dB cross-mod
1	40 dBmV	32 dBmV
2	45 dBmV	37 dBmV
3	50 dBmV	42 dBmV
4	55 dBmV	47 dBmV
5	60 dBmV	52 dBmV

New MATV systems that are designed to carry the 12 channels of a CATV feed would of necessity be engineered using the derated output capability as in the last column. The difficult job comes when you wish to upgrade an existing system to distribute these same 12 channels. Where the system under question is currently using one of the first three or four amplifiers in first column, it's a simple matter of using a larger gain amplifier with equal or higher output capability under the 12-channel rating. For example, if you found a system using a Type 2 amplifier, it could be replaced by a Type 4 amplifier. Proper operation would be achieved by running this amplifier at the previous output level of 45 dBmV per channel. This of course would be necessary since you must still provide adequate signal levels to all outlets on the system. Where you were trying to replace a Type 4 amplifier, note that the Type 5 amplifier will fall 3 dB short of meeting the system needs. Here it would pay you to test the signal level at the last outlet on the system to see if you can stand the 3 dB drop in signal level and make your judgment from there.

When confronted with larger systems that use Type 5 amplifiers or the more powerful strip amps, it will be necessary to take a different approach. Consider the partial head end output diagram in Fig. 8-1. One of the major reasons for using high level strip amps is to overcome the splitting losses encountered in creating the large number of risers needed to cover the building. If the strip head end output was 68 dBmV feeding a 16-riser system, this would mean that each riser would get a signal level input of 54 dBmV after the splitting tree. To duplicate this same capability would require use of one Type 5 amplifier for **each** feeder line as illustrated in Fig. 8-2. With a normal input signal level of 10 dBmV from the CATV drop, a preamp of some type will be needed to overcome the 14 dB of splitter loss feeding each Type 5 amplifier. This might be something like a Type 2 amp operating well within its output capability.

At first glance this appears to be a very expensive solution to the problem, and it is! But the alternatives are much less attractive unless you have an unlimited supply of free labor. To do this job with less capable equipment would mean the replacement of all the taps in the system, possibly changing the cable to a lower loss variety, and almost certainly adding a line-extending amplifier to each riser. Such line-extending

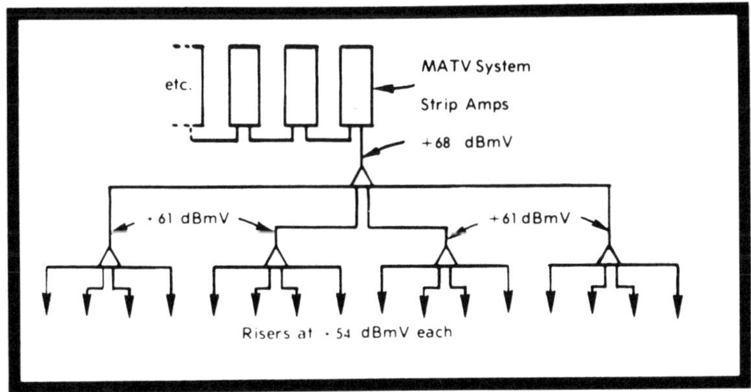

Fig. 8-1. Strip amplifier head ends deliver high level signals to each riser in the distribution system. Problem, how to revamp this system to carry CATV delivered signals.

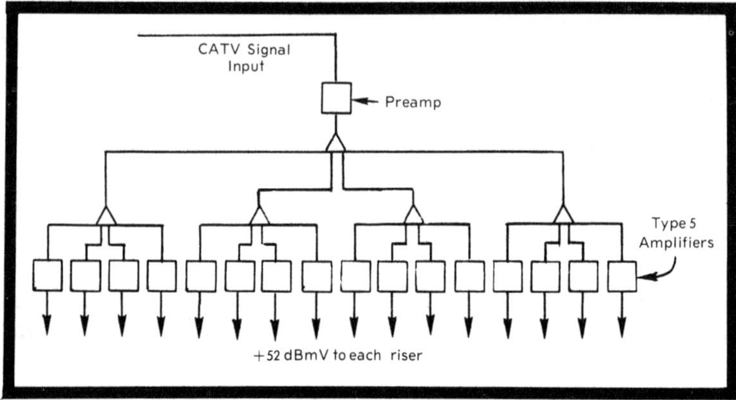

Fig. 8-2. Recommended method of implementing head end for CATV signals for system shown in Fig. 8-1. See text for discussion.

amplifiers would cost less than large ones at the head end but they would require building modification to permit mounting and bringing in of ac power to each one. All in all, the method illustrated in Fig. 8-2 turns out to be the most economical.

CONSIDERATIONS OF 30-CHANNEL CAPABILITY

The CATV cable frequency spectrum of 5 to 300 MHz was illustrated in Fig. 6-8. Allocations of all nonstandard channels are shown in the bands from 120 to 174 MHz and from 216 to 270 MHz. Use of these frequencies requires push-pull line amplifying equipment to avoid beat interference problems due to second-order distortion. Second-order distortions are the beats that can occur in any single-ended amplifier such as the second harmonic of a low band channel, the sum of two low band channels beating together, the difference between a low band channel and a high band channel, or the sum of a low band and a high band channel. All of these combinations produce beat products that land in the mid-band or super-band.

Push-pull amplifiers are really two single-ended amps in parallel in the same housing. Push-pull operation is achieved by shifting the phase of all signals in one side to be 180 deg out of phase with the other side. Both amps are carefully balanced for gain and phase across the entire band. Upon proper mixing of signals at the output, all second-order beats are canceled. A fortuitous side benefit of this design is an increase in amplifier output capability of almost 3 dB, which is sorely needed because of the increased channel loading. The jump from 12 channels to 30 channels is more than one double in the number of channels to be carried. Derating for this jump turns out to be 4 dB in order to maintain cross-mod down by the familiar minus 57 dB.

If you find it necessary to interconnect with a CATV system that is implemented for more than 12 channels, it will be necessary to use push-pull equipment in the MATV head end. Such equipment is available from CATV manufacturers and you will have to operate it within its 30 channel specs as though it were an extension to the CATV system.

PLANNING NEW MATV SYSTEMS FOR CATV

The planning of a new building distribution system that is to be fed from CATV must include some thinking about the tenants who are going to use this service. CATV implies a monthly charge for service that might be included in the rent. Premium services on CATV will undoubtedly command additional charges to the customer. The method by which service charges, both regular and premium, are imposed on the customer must be understood before the MATV system can be designed.

If service charges are to be billed directly to the tenant, you must consider the need for control of each connection. First because if a tenant falls behind on payment, you must provide a means of disconnection and, second, how will you provide TV service to the tenant who does not want the premium service offered by CATV?

The ultimate in versatility is to put **two** complete systems into the building. One system would carry the CATV signals and the other would carry the available air channels on a standard MATV system. Right away this brings visions of unacceptable double costs. It's obvious that two systems will cost more than one but it will **not** be double. The key is in **control**.

Control of each outlet can only be achieved if each outlet is served by a single cable whereby you can determine what signals are fed into each cable. To achieve this control it will be necessary to change the method of wiring a building. To keep cable costs low, you will want to find the shortest routing to a central control location for each floor or group of floors. See Fig. 8-3, which illustrates this idea.

Two head end systems are needed. One is a regular MATV head end and the other is an amplifier of the proper size to handle the CATV signals. Two main feeder cables are routed through the core of the building, out of the way of general

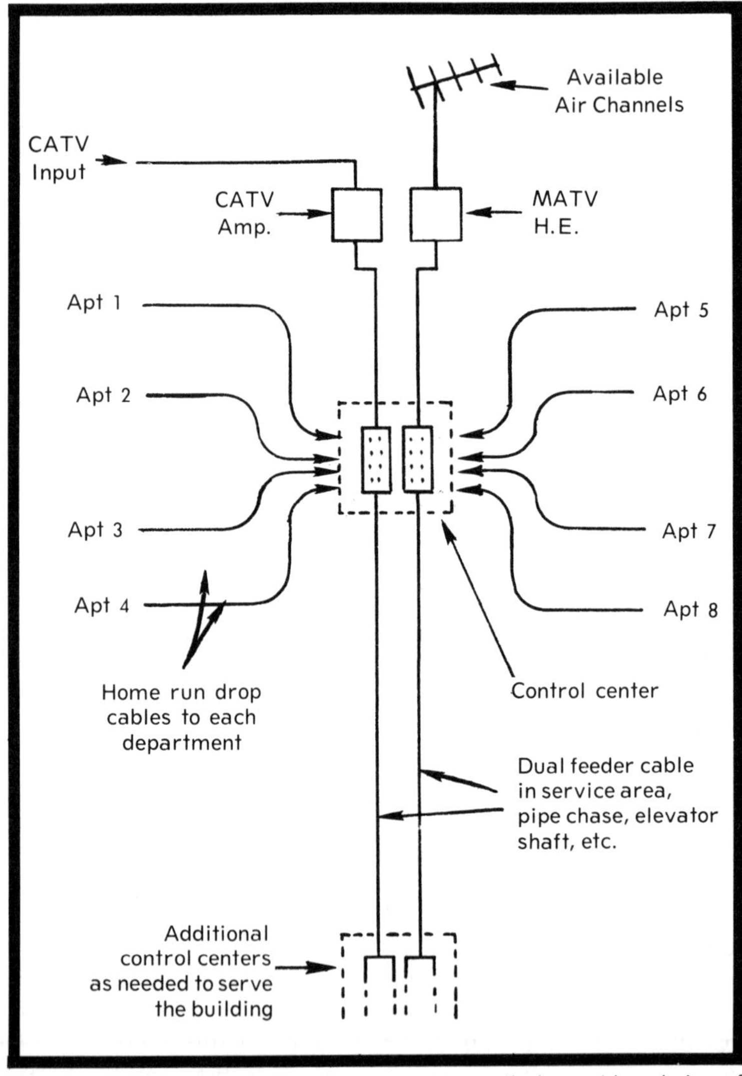

Fig. 8-3. Dual system design for new systems that provides choice of CATV or MATV service to each outlet.

public traffic. Control centers are established as necessary to service groupings of outlets. These control centers consist of multiple outlet taps on each feeder with enough taps to feed all outlets from either service. It would be a good idea to protect these taps with a lockable enclosure. Home run drop cables from each apartment run into the control center and may be connected to either TV service according to the wishes of each tenant. Within each apartment the cable would terminate in a convenient feedthrough wall outlet.

The advantages of this type of system design are in operation. Initial cost is high by comparison to a conventional MATV system but the revenue possibilities can be easily demonstrated. Following the national average for CATV subscribers, approximately 40 per cent of the tenants can be expected to want the premium TV service. This of course means extra income and should a tenant fail to pay on time, he can be disconnected without entering his apartment.

A system like this does require some special care in the selection of equipment. The biggest area for concern is the possibility of crosstalk between the two cables. The CATV cable will surely utilize the same channels that are available off the air. It will be necessary to shield each system so one does not cause interference with the other. To assure this it will be necessary to use one of the highly shielded cables used in CATV systems. This is only necessary for the feeder cables that are run together throughout the building. Drop cables should be well shielded also, but one of the foil shielded cables will suffice since there are only short runs of parallel cables in this area.

The taps used in the control centers must be well shielded also. Again the best selection is from the CATV industry. Special MATV taps may also be used if they are rated for CATV compatible service. Generally speaking they should have a radiation shielding spec of 80 dB or more. This also holds true for the head end amplifying equipment. It too should be well shielded and carry a spec of 80 dB min. All devices including taps should provide coax connectors in and out and be completely enclosed in a metal housing, preferably a casting. With the proper selection of equipment you should have a clean system free from ghosting or interference caused by the signals of one cable getting into the channels of the other cable.

TWO-WAY CATV EFFECTS ON MATV

The CATV industry is governed by a rather extensive set of technical rules. Part 74 of the FCC Rules and Regulations spells out the minimum technical requirement for CATV systems in the top 100 markets in the United States. Most of these requirements are met or exceeded by the recommendations of this book. One item is still open for discussion; that is, the rules state that all new systems after the effective date of the rules, March 1972, must be capable of two-way communications on at least a nonvoice capacity.

Note that the rules say that the systems must be **capable** of two-way. They do not say that it must actually carry any two-way information. All systems being built today are designed to include this capability, usually on a retrofit basis. This means that with the addition of plug-in equipment, such as return amps, filters, etc., these systems **could** carry two-way traffic. It's rather obvious that given the capability, many systems will implement the two-way service in one form or another. To function properly on a fully compatible basis, this two-way capability must be considered in MATV engineering.

The basic engineering principles as discussed in this book do not change just because of two-way. The things that do change are the equipment that is selected for use in the MATV system that is to be connected to CATV. To appreciate the differences between equipment needs in the past and the needs going forward, let us look at a few specs.

Frequency Spectrum

The frequency spectrum considered as standard for CATV is 5 MHz to 300 MHz. To have a fully compatible system with this spec it is necessary to have passive devices meet this frequency range. This involves such devices as splitters, directional couplers, and taps. Most splitters and directional couplers have some response below 54 MHz but many will quit functioning in the 10 MHz to 20 MHz range. Taps vary all over the place. Any tap that is not specified to cover down to 5 MHz should be suspect. Some purely resistive taps will work well below 5 MHz while others, especially "tilted" isolation taps, are sure to give trouble at frequencies below Ch. 2. Care must also be given for those frequencies above Ch. 13. Here, taps will produce the most problems. The low values of isolation are usually tilted and the insertion losses around 300 MHz could be a real problem.

"Tilted" refers to the isolation value of a tap. "Tilted" taps have high isolation at low frequencies and low isolation at high frequencies. Such configurations are an attempt to compensate for the reverse effect in cable, and produces an effect similar to an equalizer.

Return Signals

The nature of return signals are very much in the formative stages and will probably stay there for a number of years. Regardless of what form the return signals may take, it is incumbent on the system equipment to get those signals from the tap back to the head end. This requires some attention in tap design. The best tap circuitry for this purpose is the directional coupler. With directional coupler taps, all the energy that is fed back into the tap will be directed toward the head end and none will be wasted in the termination at the end of the feeder line.

Interference to Return Signals

The return spectrum is in the band of frequencies from 5 MHz to 30 MHz. This is an area where MATV engineers have not been concerned except for infrequent subchannel jobs. Even in subchannel work, interference in this band has been minimal. This is not the case when considering a system the size of CATV. In MATV work there may be a few hundred TV sets connected. In a CATV system, there will be many thousands of TV sets connected and herein lies the problem.

Within every TV set there are many high energy signals of complicated waveform needed to make the TV set do its thing. Of prime concern is the horizontal sweep at approximately 15,750 Hz. It is a complex wave that contains many harmonics. Some of these harmonics can appear above 5 MHz. Other combinations of signals in the tuners of TV sets can produce heretofore ignored, spurious signals in the return spectrum. Some of these signals will find their way to the antenna terminals of the TV set and will feed back over the cable drop as return signal energy. The contribution of any one TV set will surely be small but when you consider a few thousand sets, contributing a few microvolts each, the return spectrum could easily become filled with interference and be rendered unusable.

This also applies to MATV systems that are connected to the CATV system. A means of blocking these unwanted signals must be built into every system. Where a tenant has subscribed to the full services of the CATV system including the extra channels in the mid- and super-bands, there will be a converter between his TV set and the incoming cable. This converter will provide all the blocking of unwanted signals from the TV set that is needed. Some subscribers will be connected directly to the system without a converter because they are content with what the regular channels (2 through 13) have to offer. They won't get a converter and **it's this type of connection that you must worry about**.

To cope with this problem there are high pass filters available to block the 5 MHz to 30 MHz band. The easiest way to keep this band clean is to use one of the matching transformers that has the high pass filter built in. These transformers cost a little more but are a must in active two-way systems.

CONCLUSION

MATV systems engineering is a multifaceted business as has been discussed throughout this book. MATV systems cover the gamut of multiple outlets for a home to very large systems that are part of even larger CATV systems. The challenge of this business is in being able to cope with its many and varied demands. I trust that in some small way this book will aid you in what I have found to be an exciting and rewarding vocation.

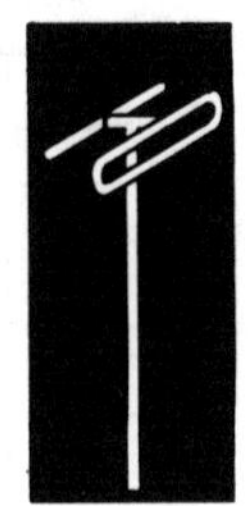

Appendix

Tables, Graphs, and Formulas

Table 1. Conversion from decibels to microvolts

dB	μV	dB	μV	dB	μV
– 40	10.00	0	1 000	41	112 200
– 39	11.22	1	1 122	42	125 900
– 38	12.59	2	1 259	43	141 300
– 37	14.13	3	1 413	44	158 500
– 36	15.85	4	1 585	45	177 800
– 35	17.78	5	1 778	46	199 500
– 34	19.95	6	1 995	47	223 900
– 33	22.39	7	2 239	48	251 200
– 32	25.12	8	2 512	49	281 800
– 31	28.18	9	2 818	50	316 200
– 30	31.62	10	3 162	51	354 800
– 29	35.48	11	3 548	52	398 100
– 28	39.81	12	3 981	53	446 700
– 27	44.67	13	4 467	54	501 200
– 26	50.12	14	5 012	55	562 300
– 25	56.23	15	5 623	56	631 000
– 24	63.10	16	6 310	57	707 900
– 23	70.79	17	7 079	58	794 300
– 22	79.43	18	7 943	59	891 300
– 21	89.13	19	8 913	60	1 000 000
– 20	100.0	20	10 000	61	1 122 000
– 19	112.2	21	11 220	62	1 259 000
– 18	125.9	22	12 590	63	1 413 000
– 17	141.3	23	14 130	64	1 585 000
– 16	158.5	24	15 850	65	1 778 000
– 15	177.8	25	17 780	66	1 995 000
– 14	199.5	26	19 950	67	2 239 000
– 13	223.9	27	22 390	68	2 512 000
– 12	251.2	28	25 120	69	2 818 000
– 11	281.8	29	28 180	70	3 162 000
– 10	316.2	30	31 620	71	3 548 000
– 9	354.8	31	35 480	72	3 981 000
– 8	398.1	32	39 810	73	4 467 000
– 7	446.7	33	44 670	74	5 012 000
– 6	501.2	34	50 120	75	5 623 000
– 5	562.3	35	56 230	76	6 310 000
– 4	631.0	36	63 100	77	7 079 000
– 3	707.9	37	70 790	78	7 943 000
– 2	794.3	38	79 430	79	8 913 000
– 1	891.3	39	89 130	80	10 000 000
– 0	1000.0	40	100 000		

Table 2. Conversion from decibels to voltage multiplier.

dB	Voltage Multiplier	dB	Voltage Multiplier	dB	Voltage Multiplier
1	1.12	16	6.3	31	35
2	1.25	17	7	32	40
3	1.4	18	8	33	45
4	1.6	19	9	34	50
5	1.8	20	10	35	56
6	2	21	11	36	63
7	2.25	22	12.5	37	71
8	2.5	23	14	38	80
9	2.75	24	16	39	90
10	3.16	25	18	40	100
11	3.55	26	20	43	140
12	4	27	22.5	46	200
13	4.5	28	25	50	300
14	5	29	28	56	600
15	5.6	30	32	60	1000

Table 3. TV Channel Assignments

Ch	Freq (MHz)	½ λ in.	Pix	Color	Sound
T-7	5.75-11.75		7	10.58	11.5
T-8	11.75-17.75		13	16.58	17.5
T-9	17.75-23.75		19	22.58	23.5
T-10	23.75-29.75		23	28.58	29.5
T-11	29.75-35.75		31	34.58	35.5
T-12	35.75-41.75		37	40.58	41.5
T-13	41.75-47.75		43	46.58	47.5
VHF LOW BAND					
2	54-60	103.8	55.25	58.83	59.75
3	60-66	93.8	61.25	64.83	65.75
4	66-72	85.7	67.25	70.83	71.75
5	76-82	74.8	77.25	80.83	81.75
6	82-88	69.5	83.25	86.83	87.75
FM	88-108	—	—	—	—
VHF HIGH BAND					
7	174-180	33.4	175.25	178.83	179.75
8	180-186	32.8	181.25	184.83	185.75
9	186-192	31.5	187.25	190.83	191.75
10	192-198	30.3	193.25	196.83	197.75
11	198-204	29.4	199.25	202.83	203.75
12	204-210	28.5	205.25	208.83	209.75
13	210-216	27.7	211.25	214.83	215.75
UHF CHANNELS					
14	470-476	12.5	471.25	474.83	475.75
15	476-482	12.4	477.25	480.83	481.75
16	482-488	12.2	483.25	486.83	487.75
17	488-494	12.0	489.25	492.83	493.75
18	494-500	11.9	495.25	498.83	499.75
19	500-506	11.8	501.25	504.83	505.75
20	506-512	11.6	507.25	510.83	511.75
21	512-518	11.5	513.25	516.83	517.75
22	518-524	11.3	519.25	522.83	523.75
23	524-530	11.2	525.25	528.83	529.75
24	530-536	11.1	531.25	534.83	535.75
25	536-542	11.0	537.25	540.83	541.75
26	542-548	10.8	543.25	546.83	547.75
27	548-554	10.7	549.25	552.83	553.75
28	554-560	10.6	555.25	558.83	559.75
29	560-566	10.5	561.25	564.83	565.75
30	566-572	10.4	567.25	570.83	571.75
31	572-578	10.3	573.25	576.83	577.75
32	578-584	10.2	579.25	582.83	583.75
33	584-590	10.1	585.25	588.83	589.75
34	590-596	10.0	591.25	594.83	595.75
35	596-602	9.9	597.25	600.83	601.75
36	602-608	9.8	603.25	606.83	607.75
37	608-614	9.7	609.25	612.83	613.75
38	614-620	9.6	615.25	618.83	619.75
39	620-626	9.5	621.25	624.83	625.75
40	626-632	9.4	627.25	630.83	631.75
41	632-638	9.3	633.25	636.83	637.75
42	638-644	9.2	639.25	642.83	643.75
43	644-650	9.2	645.25	648.83	649.75
44	650-656	9.1	651.25	654.83	655.75
45	656-662	9.0	657.25	660.83	661.75
46	662-668	8.9	663.25	666.83	667.75
47	668-674	8.8	669.25	672.83	673.75
48	674-680	8.7	675.25	678.83	679.75
49	680-686	8.7	681.25	684.83	685.75
50	686-692	8.6	687.25	690.83	691.75
51	692-698	8.5	693.25	696.83	697.75
52	698-704	8.4	699.25	702.83	703.75
53	704-710	8.4	705.25	708.83	709.75
54	710-716	8.3	711.25	714.83	715.75
55	716-722	8.2	717.25	720.83	721.75
56	722-728	8.1	723.25	726.83	727.75
57	728-734	8.0	729.25	732.83	733.75
58	734-740	8.0	735.25	738.83	739.75
59	740-746	7.9	741.25	744.83	745.75
60	746-752	7.8	747.25	750.83	751.75
61	752-758	7.8	753.25	756.83	757.75
62	758-764	7.7	759.25	762.83	763.75
63	764-770	7.7	765.25	768.83	769.75
64	770-776	7.6	771.25	774.83	775.75
65	776-782	7.5	777.25	780.83	781.75
66	782-788	7.5	783.25	786.83	787.75
67	788-794	7.4	789.25	792.83	793.75
68	794-800	7.4	795.25	798.83	799.75
69	800-806	7.3	801.25	804.83	805.75
70	806-812	7.3	807.25	810.83	811.75
TRANSLATOR CHANNELS					
71	812-818	7.2	813.25	816.83	817.75
72	818-824	7.2	819.25	822.83	823.75
73	824-830	7.1	825.25	828.83	829.75
74	830-836	7.0	831.25	834.83	835.75
75	836-842	7.0	837.25	840.83	841.75
76	842-848	6.9	843.25	846.83	847.75
77	848-854	6.9	849.25	852.83	853.75
78	854-860	6.8	855.25	858.83	859.75
79	860-866	6.8	861.25	864.83	865.75
80	[illegible]	[illegible]	[illegible]	870.83	871.75
81	872-878	6.7	873.25	876.83	877.75
82	878-884	6.7	879.25	892.83	883.75
83	884-890	6.6	885.25	888.83	889.75

Table 4. Voltage standing wave ratio (VSWR) to decibel match.

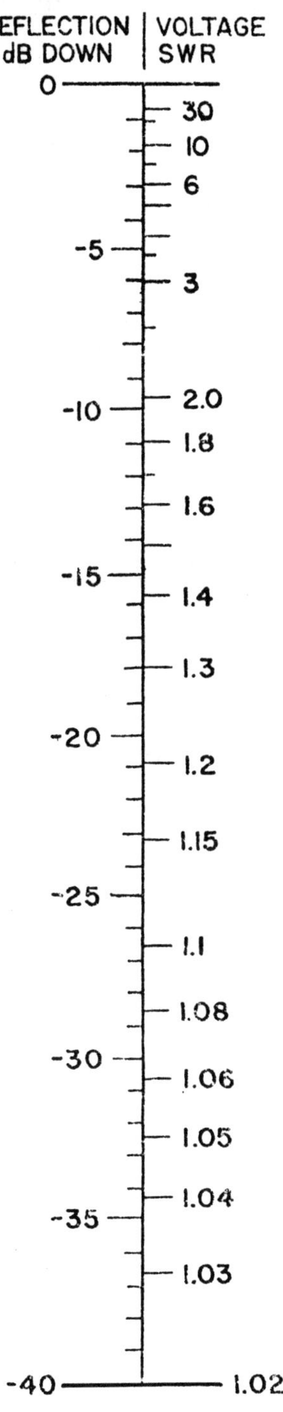

Table 5. Permissible UHF to VHF conversions.

	UHF CONVERSIONS						VHF CONVERSIONS						
Ch		2	3	4	5	6	7	8	9	10	11	12	13
	Bottom	54	60	66	76	82	174	180	186	192	198	204	210
14	470	416	410	404	*	388	296	290	*	278	272	266	260
15	476	422	416	410	*	394	302	296	290	284	278	272	266
16	482	428	422	416	*	400	308	302	296	*	284	278	272
17	488	434	428	422	412	406	314	308	302	*	290	284	278
18	494	440	434	428	418	*	320	314	308	302	296	290	284
19	500	446	440	434	424	*	326	320	314	308	*	296	290
20	506	452	446	440	430	*	332	326	320	314	308	302	296
21	512	458	452	446	436	*	338	332	326	320	314	*	302
22	518	464	458	452	442	*	*	338	332	326	320	*	308
23	524	470	464	458	448	442	*	344	338	332	326	320	314
24	530	476	470	464	454	448	*	350	344	338	332	326	*
25	536	482	476	470	460	454	*	*	350	344	338	332	326
26	542	488	482	476	466	460	368	*	356	350	344	338	332
27	548	494	488	482	472	466	374	*	362	356	350	344	338
28	554	500	494	488	478	472	380	*	*	362	356	350	344
29	560	506	500	494	484	478	386	380	*	368	362	356	350
30	566	512	506	500	490	484	392	386	*	374	368	362	356
31	572	518	512	506	496	490	398	392	*	*	374	368	362
32	578	524	518	512	502	496	404	398	392	*	380	374	368
33	584	530	524	518	508	502	410	404	398	*	386	380	374
34	590	536	530	524	514	508	416	410	404	*	*	386	380
35	596	542	536	530	520	514	422	416	410	404	*	392	386
36	602	548	542	536	526	520	428	422	416	410	*	398	392
37	608	554	548	542	532	526	434	428	422	416	*	*	398
38	614	560	554	548	538	532	440	434	428	422	416	*	404
39	620	566	560	554	544	538	446	440	434	428	422	*	410
40	626	572	566	560	550	544	452	446	440	434	428	*	*
41	632	578	572	566	556	550	458	452	446	440	434	428	*
42	638	584	578	572	562	556	464	458	452	446	440	434	*
43	644	590	584	578	568	562	470	464	458	452	446	440	*
44	650	596	590	584	574	568	476	470	464	458	452	446	440
45	656	602	596	590	580	574	482	476	470	464	458	452	446
46	662	608	602	596	586	580	488	482	476	470	464	458	452
47	668	614	608	602	592	586	494	488	482	476	470	464	458
48	674	620	614	608	598	592	500	494	488	482	476	470	464
49	680	626	620	614	604	598	506	500	494	488	482	476	470
50	686	632	626	620	610	604	512	506	500	494	488	482	476
51	692	638	632	626	616	610	*	512	506	500	494	488	482
52	698	644	638	632	622	616	*	518	512	506	500	494	488
53	704	650	644	638	628	622	*	524	518	512	506	500	494
54	710	656	650	644	634	628	*	530	524	518	512	506	500
55	716	662	656	650	640	634	*	*	530	524	518	512	506
56	722	668	662	656	646	640	548	*	536	530	524	518	512
57	728	674	668	662	652	646	554	*	542	536	530	524	518
58	734	680	674	668	658	652	560	*	548	542	536	530	524
59	740	686	680	674	664	658	566	*	*	548	542	536	530
60	746	692	686	680	670	664	572	566	*	554	548	542	536
61	752	698	692	686	676	670	578	572	*	560	554	548	542
62	758	704	698	692	682	676	584	578	*	566	560	554	548
63	764	710	704	698	688	682	590	584	*	*	566	560	554
64	770	716	710	704	694	688	596	590	584	*	572	566	560
65	776	722	716	710	700	694	602	596	590	*	578	572	566
66	782	728	722	716	706	700	608	602	596	*	584	578	572
67	788	734	728	722	712	706	614	608	602	*	*	584	578
68	794	740	734	728	718	712	620	614	608	602	*	590	584
69	800	746	740	734	724	718	626	620	614	608	*	596	590
70	806	752	746	740	730	724	632	626	620	614	*	602	596
71	812	758	752	746	736	730	638	632	626	620	*	*	602
72	818	764	758	752	742	736	644	638	632	626	620	*	608
73	824	770	764	758	748	742	650	644	638	632	626	*	614
74	830	776	770	764	754	748	656	650	644	638	632	*	620
75	836	782	776	770	760	754	662	656	650	644	638	*	*
76	842	788	782	776	766	760	668	662	656	650	644	638	*
77	848	794	788	782	772	766	674	668	662	656	650	644	*
78	854	800	794	788	778	772	680	674	668	662	656	650	*
79	860	806	800	794	784	778	686	680	674	668	662	656	*
80	866	812	806	800	790	784	692	686	680	674	668	662	656
81	872	818	812	806	796	790	*	692	686	680	674	668	662
82	878	824	818	812	802	796	*	698	692	686	680	674	668
83	884	830	824	818	808	802	*	704	698	692	686	680	674

Table 6. Loss vs frequency for various cables.

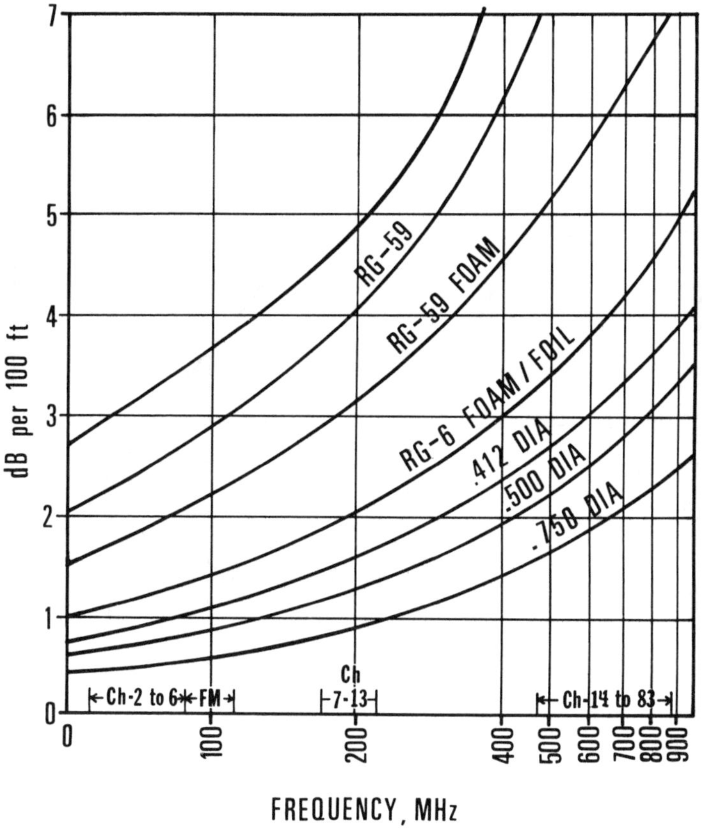

Table 7. Nomograph for adding unequal levels of Voltage (A) or Power (B).

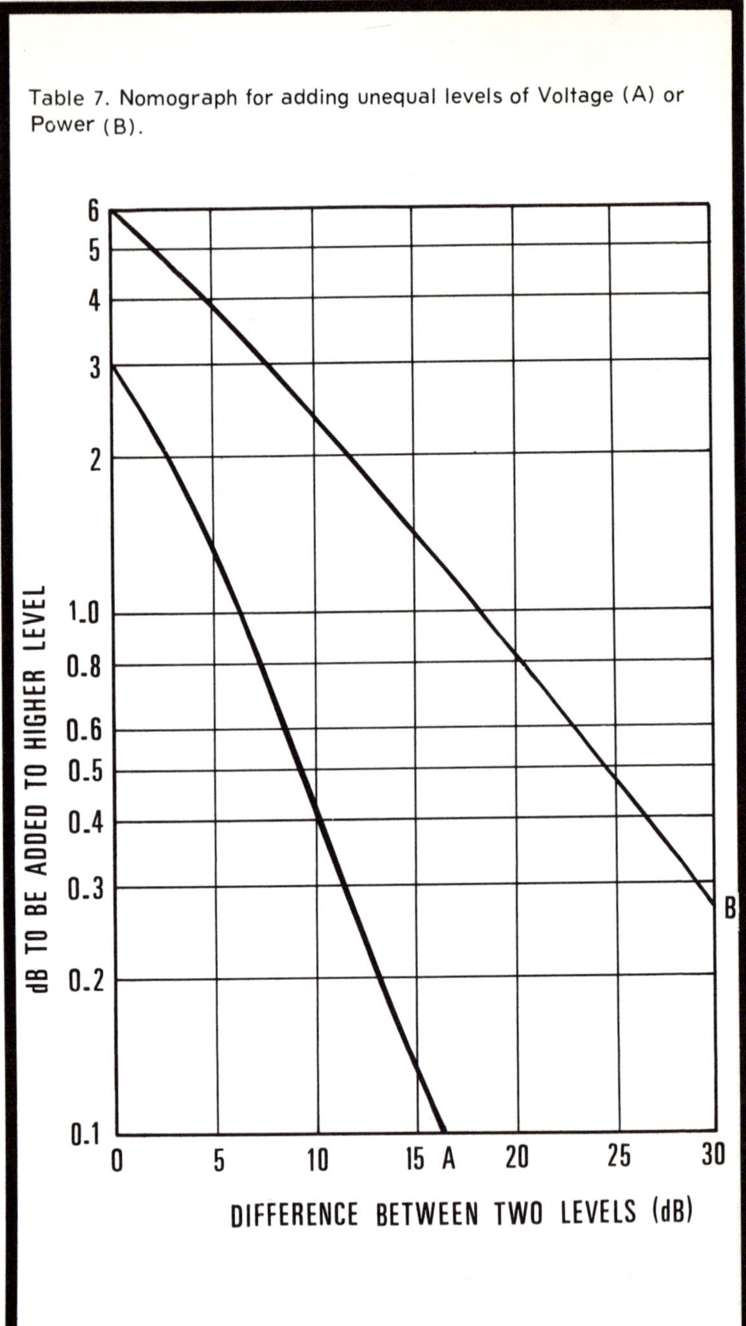

70868

SOUTH HILLS ELECTRONICS COMPANY

1936 W. Liberty Ave. Pittsburgh, Pa. 15226

Phone: 341-6200

BRANCH STORE

201 Anderson St. Pittsburgh, Pa. 15212

Phone: 321-5400

Date 3-23- 19 78

Sold To ADAMS FURNITURE.

Address

1	657 TAB.	7	95
1	G.E. TRANS VOH. book.	—	

TAX EXEMPT.

3-23-78

Thank You!

509 MOORE BUSINESS FORMS, INC., L

Index

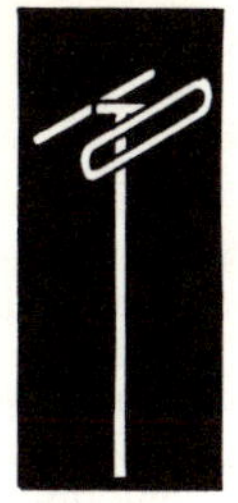

A

B

C

D

E

F